Energy Taxes
and Subsidies

A Report to the Energy Policy Project of the Ford Foundation

Energy Taxes and Subsidies

Gerard M. Brannon

Ballinger Publishing Company ● Cambridge, Mass.
A Subsidiary of J.B. Lippincott Company

Published in the United States of America by Ballinger Publishing Company
Cambridge, Mass.

First Printing, 1974

Library of Congress Catalog Card Number: 74-9645

International Standard Book Number: 0-88410-308-0 HB
                                      0-88410-329-3 PB
Printed in the United States of America

Library of Congress Cataloging in Publication Data

Brannon, Gerard Marion, 1922–
         Energy taxes and subsidies.

         1. Energy policy–United States. 2. Power resources–
Taxation–United States. 3. Subsidies–United States.
4. Public utilities–Rates. I. Title.
HD9502.U52B68   333.7'0973   74-9645
ISBN   0-88410-308-0
        0-88410-329-3

Table of Contents

List of Tables

Tables

Figure

Foreword

In December 1971 the Trustees of the Ford Foundation authorized the organization of the Energy Policy Project. In subsequent decisions the Trustees have approved supporting appropriations to a total of $4 million, which is being spent over a three-year period for a series of studies and reports by responsible authorities in a wide range of fields. The Project Director is S. David Freeman, and the Project has had the continuing advice of a distinguished Advisory Board chaired by Gilbert White.

This analysis of "Energy Taxes and Subsidies" is one of the outputs of the Project. As Mr. Freeman explains in his Preface, neither the Foundation nor the Project presumes to judge the specific conclusions and recommendations of the Committee which prepared this volume. We do commend this report to the public as a serious and responsible analysis which has been subjected to review by a number of qualified readers.

This study, perhaps even more than others in the Project, deals with sensitive and difficult questions of public policy. Not all of it is or could be easy reading; and not all those we have consulted have agreed or could be expected to agree with all of it. Taxes and subsidies are among the most powerful tools available to government in shaping the nation's energy and economic future. Decisions to establish, change or retain various tax provisions or subsidy programs can have profound impacts on energy prices, growth rates, production and consumption patterns, income levels and distributions and international trade patterns. No single study can exhaust a subject which is so complex, controversial, and partly obscured by gaps in available data and understanding. And different readers inevitably will have different perspectives.

In this last respect the present study reflects tensions which are intrinsic to the whole of the Energy Policy Project—tensions between one set of objectives and another. As the worldwide energy crisis has become evident to us all, we have had many graphic illustrations of such tensions, and there are more ahead. This is what usually happens when a society faces hard choices, all of them carrying costs that are both human and material.

But it is important to understand that there is a fundamental difference between present tension and permanent conflict. The thesis accepted by our Board of Trustees when it authorized the Energy Policy Project was that the very existence of tension, along with the inescapable necessity for hard choices, argued in favor of studies which would be, as far as possible, fair, responsible, carefully reviewed, and dedicated only to the public interest. We do not suppose that we can evoke universal and instantaneous agreement, and still less do we presume that this Project can find all the answers. We do believe that it can make a useful contribution to a reasonable and democratic resolution of these great public questions, one which will serve the general interest of all.

This study on energy tax and subsidy policies is an excellent example of the kind of issue-clarifying and thought-provoking study we aim at. It combines strong arguments for market-oriented energy policies with specific tax and subsidy policy recommendations which, in the author's considered and thoughtful opinion, will make markets more economically efficient allocators of society's scarce resources. The reader may, as several of our reviewers have, disagree with the author's stated goal of economic efficiency, or with his assessment of current conditions, or with his specific recommendations. But careful reading and interchange of views should clarify the areas of disagreement about facts or values, thereby raising the level of public discussion of these critical public policy issues.

McGeorge Bundy
President, Ford Foundation

Preface

The Energy Policy Project was initiated by the Ford Foundation in 1971 to explore alternative national energy policies. This book, *Energy Taxes and Subsidies,* is one of the series of studies commissioned by the Project. It is presented here as a carefully prepared contribution by the author to today's public discussion about the taxation of the energy industry. It is our hope that each of these special reports will stimulate further thinking and questioning in the specific areas that it addresses. The special report are being released as they are completed rather than delaying their release until the final report of the Energy Policy Project is completed because I believe they can make a timely contribution to the public discussion of energy policies. At the very most, however, each special report deals with only a part of the energy puzzle; our final report, to be published later in 1974, will attempt to integrate these parts into a comprehensible whole, setting forth the energy policy options available to the nation as we see them.

This book, like the others in the series, has been reviewed by scholars and experts in the field not otherwise associated with the Project in order to be sure that differing points of view were considered. With each book in the series, we offer reviewers the opportunity of having their comments published in an appendix, but none chose to do so with this volume.

Energy Taxes And Subsidies is the author's report to the Ford Foundation's Energy Policy Project and neither the Foundation, its Energy Policy Project or the Project's Advisory Board have assumed the role of passing judgement on its contents or conclusions. We will express our views in the Project's final report that will complete this series of publications.

S. David Freeman
Director
Energy Policy Project

Acknowledgments

This book is one of a series of background studies financed by the Energy Policy Project of the Ford Foundation. The design of the work resulted from specifications of the staff of the Project on questions in the tax and subsidy area that needed further research in the development of its own final report. The work also grew out of a division of labor between the tax-subsidy researchers and several teams who were simultaneously doing other background studies making up this series.

This study benefited from discussions with the staff of the Energy Policy Project. The contribution of Walter Mead was monumental. Valuable help was obtained from David Freeman, William lulo, Kenneth Saulter, and Monte Canfield. Several consultants to the Energy Policy Project contributed valuable advice in the early design, particularly Arnold Hargerber, James Buchanan, and David and Attiat Ott. Consultation with Philip Verleger of Data Resources, Inc., because of his involvement with various energy models being developed for the Energy Policy Project, was also valuable in designing this study. The work is, however, a report to the Energy Policy Project. The conclusions are ours and not necessarily theirs.

The final form of this book also benefited from the advice of a review committee and a symposium held August 7, 1973, in which a first draft was reviewed.

The body of this work depended critically on a team of researchers who have written detailed essays dealing with the subject matter of most of the chapters of this work. This team was made up of Allen Manvel (the state and local part of Chapter Two), Joseph Stiglitz (Chapter Three), Frederick Peterson (also Chapter Three), Paul Davidson, Laurence Falk, and Hoesung Lee (Chapter Four and Twelve), James Cox and Arthur Wright (Chapter Five), Glenn Jenkins (Chapter Six), Robert Spann (Chapter Seven), James Griffin (Chapter Eight), Bruce Davey and Bruce Duncombe (Chapter Ten) and John Tucillo (Chapter Twelve). With the exception of the work of Spann and Tucillo, these essays will be published in a separate volume in this series, *Studies in Energy Tax Policy*. The Spann essay will be published in another volume in this series dealing with public utility regulation.

I want particularly to acknowledge the helpful comments on the main report by Boris Bittker, James Buchanan, Paul Davidson, Stephen McDonald, William McDonnell, Richard Musgrave, Joseph Pechman, Stanley Surrey, and Arthur Wright. Two very thorough reviews by Professor Surrey leave me much in his debt. Valuable assistance on the environmental section was obtained from James Giffin and Lawrence Moss. Special ackowledgement is due to Leslie Cookenboo of Exxon, Paul Little of Mobil Oil, Minor Jameson of the Independent Petroleum Producers Association, Claude Dodgen of Texas Pacific, and Robert Stauffer of the National Coal Association; these gentlemen gave me important industry insights even though they each may have serious reservations about some of my conclusions. I have benefited from conversations with Helmut Franck and from his making available to me his drafts of work on a similar problem. I have benefited also from conversations on the foreign investment tax provisions with Thomas Field, Ira Tannenbaum, and Gary Hufbauer. An enormous debt of gratitude is owed to Kristina Goodnough, who contributed to this work in many ways as research assistant, typist, and proofreader, to Lynda Frank who did a thorough editing of the final manuscript, and to Elizabeth McKinney who turned the marked-up pages into readable copy.

Finally, I owe apologies to Frances, Margie, Paul, and Rich who patiently tolerated many dull evenings while I was holed up in the back room.

With this kind of advice and help, perfection should have been attainable. Failure to reach it lies solely with myself.

Gerard M. Brannon
Georgetown University

Energy Taxes
and Subsidies

Chapter One

The Role of Taxes and Subsidies in United States Energy Policy

1.1 THE ENERGY PROBLEM

When the work leading to this volume began in early 1973, "the energy crisis" was the current name for a set of long-term problems of energy supply and demand. Even then there had been some local shortages of heating oil, natural gas, and gasoline, but most of the "horrors" of the energy crisis were in the future.

When the final revision of the text began in early 1974, "the energy crisis" was the name for a very specific set of problems that were triggered by the Arab embargo and, more importantly, by the OPEC (Organization of Petroleum Exporting Countries) price increase announced during the Yom Kippur war. The issues that could be examined rather leisurely in early 1973 had by early 1974 been converted into such questions as, "Do we need to start gas rationing next month?"

Despite appearances, these two versions of the energy crisis were very closely related. Elements of the long-run concerns of early 1973 were the prospect of the United States being cut off from Arab oil sources and our susceptibility to higher world oil prices. These developments came suddenly in late 1973 on top of growing demand and lagging domestic supply, elements of the long-term problem that was foreseen in early 1973.

We have been operating under some degree of general price control for over two years and under natural gas price control for 19 years. Until 1973 we limited oil imports, and it should have surprised no one that there were neither spare tankers nor spare refining capacity to handle more crude when we suddenly wanted to expand imports in the spring of 1973. In the past four years we have begun to control a number of environmental pollution problems related to energy that we barely knew about ten years ago. The growth in nuclear generation of electricity that we planned for has been much slower than expected. Environmental pressures have slowed down refinery construction and public utility expansion through delays in site approval and have post-poned the delivery of oil from Alaska. While all this has been going on, new cars have been using up more gasoline per mile due to a combination of factors—preferences for such things as air-conditioners and power accessories and enforcement of environmental controls on motor design.[1]

When any new problem settles down to the point of taking definite form, our economy can begin to adjust to it. Energy problems, however, have been uniquely bothersome because they have mostly taken the form of rapidly growing demands and slow-growing or lessen-ing supplies, both of which are the normal economic signals for rising prices.

Energy production is heavily dependent on limited resources, so that using up the known reserves faster means an increasingly costly search for more reserves or turning to more expensive substitutes. In addition, the reasonable objective of finding less polluting forms of energy production means more costly energy, and the oil-producing countries are devoted to raising prices. As prices rise, most sellers of energy will get richer and most buyers of energy will get poorer.

It overstates the role of public policy to say that there must be policy on how the United States will respond to new facts about energy. Most of the response will occur as a result of individuals reassessing their energy "needs" in the light of higher prices and as a result of businesses exploring new energy-producing combinations that may become manageable in the light of the same higher prices.

Rather, public policy must be concerned with how efficiently the market will respond to price signals and with how fair the income changes brought about the the prices and new supplies will be. Will the best adjustments come about? Will there be loss of jobs? Will the price changes involve exorbitant gains for some and brutal burdens for others? How can the process of adjustment be improved in the short run and in the long run?

1.2 TAXES AND SUBSIDIES—PRICES AND INCOMES

Other volumes in this series deal with many sides of the complex energy problem and the policies needed to assure that prospective changes in the energy industry are efficient and fair. This book is about the ways government uses, or could use, taxes and subsidies in its energy policies.

Taxes and subsidies are similar government energy policies because both are interventions in the working price system, interventions that direct the energy industry and determine the respective incomes of capital, land, and labor that make up the energy industry.

The word "intervention" acknowledges that taxes and subsidies are involved in the price system for better or for worse. In some cases people think the goal of tax policy is to meet the revenue needs of government with as little distortion of prices as possible. In other cases people advocate taxes or subsidies to offset some defect in the price system. Whether one wants to increase or decrease intervention, it is important to say something at the start about the price system in which the tax-subsidy interventions occur.

In a market economy the price system has a critical role in handling changes. One change of the kind involved in the energy problem is that petroleum may become more difficult to find—to the point that people begin to talk about a potential "shortage" of such petroleum products as gasoline or heating oil.

If it were operating ideally, the price system would prevent such problems as petroleum shortages from becoming public policy problems. If there were not enough petroleum to meet all demand at current prices, there would be temporary shortages that would lead to higher prices that in turn would (1) discourage some potential buyers and (2) encourage more potential producers. The increase in price would continue until the shortages that activated it are eliminated; then prices would stop increasing. In fact, in a good price system, with business planning ahead, corrections are made before shortages actually occur.

Ideally, the price system should do the job of organizing production and consumption better than government planners.[2] The temporary oil shortage of our example could have been corrected by less demand or by more supply. The role of these two responses is at root too complex for government planners. How much demand should fall off depends on how important the advantages of oil as an energy source are to consumers. (Oil is more important for some than for others.) How much supply should rise depends on the cost of finding and producing more oil or more oil substitutes. With a price system operating properly, these complex technical questions would be continuously answered by decisions of producers and consumers faced with specific current prices. We would not need government to tell consumers to use less oil or to tell producers how much more oil to produce.

Even if the supply of petroleum under the ground were severely limited, a properly functioning price system would forestall an oil shortage. In this case a rising price would primarily discourage potential buyers, thereby efficiently rationing the limited supply to those least able to utilize alternative energy sources.

In Figure 1–1 we have a picture of demand and supply. The demand picture is represented by the number of empty barrels that consumers are willing to pay to have filled up. At a price of $6 a barrel they are willing to buy 14,000,000 barrels a day (14 mb/d); at a price of $3 they would buy 22 mb/d. The lower solid line is the old supply situation, which might be thought of as the possible oil flow. That line tells us that at a price of $3 the producers would be willing to fill up 15 mb/d; at $4.25, 20 mb/d. At a price of $4 the market is cleared; producers will supply 19 mb/d, which is exactly what consumers are willing to pay for at that price.

The upper solid line is a changed supply situation, which might be due to the disappearance of foreign oil or to exhaustion of the best supplies. Either situation would cause more of the oil to be drawn from deeper wells or offshore wells. The new supply picture tells us that at a price of $4 per barrel producers will be willing to sell only 13 mb/d, and at a price of $6, 19 mb/d. Because we have not changed the demand picture, the market clearing now occurs at a price of $5 and production of 16.5 mb/d.

The usual problem with raising prices—that you lose some buyers—will not apply for a while because at the old price and the new supply there will be just too many buyers. When the price gets to $5, the amounts demanded and supplied will be back in balance.

Very clearly, Americans are sometimes dissatisfied with relying on higher prices to bring about the coordination of the quantity demanded and the quantity supplied. This has led us to impose price ceilings by law and, when we have had persistent shortages, to limit demand by rationing. In the picture of demand and supply in Figure 1–1, we can point out the effects of a price ceiling of $4 a barrel. At this point there would be demand for 19 mb/d but a supply of only 13mb/d, which would mean a lot of dissatisfied customers, some of whom would be willing to pay black market prices. To prevent black markets from undercutting the price ceiling, steps must be taken to cut back the demand—by patriotic appeals, by closing gas stations on Sunday, by rationing gasoline, and the like.

Preventing price increases is not an easy path to follow.

1.3   DEFECTS IN ENERGY PRICES

We have taken pains to clarify what a good price system should do, with full awareness that price systems in the real world are imperfect. People do sometimes prefer the complications of price control and rationing to the market price result. Why? What is wrong with the market price outcome?

It is *not* one of the defects of the real world that people do things for profit. When the price system is working properly, it is profitable for business to produce the goods and services that customers want.

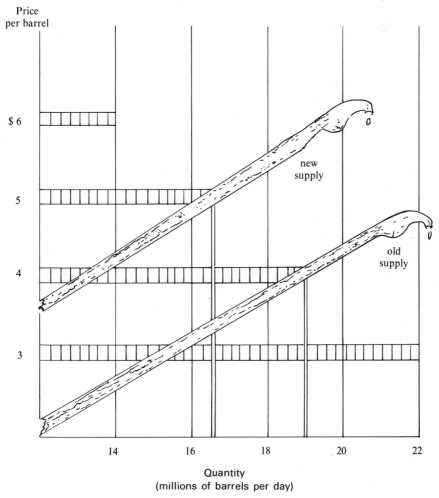

Figure 1.1. How Price Relates the Quantity Demanded to the Quantity Supplied

The way an ideal price system accomplishes its coordinating role is that the price at which various products and services sell determines how much people can earn from producing them. People who are able to offer services of labor and capital and land try to sell them to the highest bidder, and what a business can bid for productive labor, land, and capital depends on the price at which the products can be sold.

A century ago the price of anthracite coal made it possible for companies to hire capital, labor, and engineering skills to open new anthracite mines. Current prices, on the contrary, send out the message that the public prefers energy in different forms, such as oil. Businesses were organized to produce more oil because this was more profitable than opening new anthracite mines.[3]

There is a defect in the price system, however, if the wrong signals go out. It can happen that the wrong changes occur in the energy industry. Instead of shortages being prevented, we get recurrent energy shortages or excessive price increases or more pollution.

There are many ways that price systems can convey bad information and thus yield bad results. Some of the common defects are described by economists as "externalities."[4] This term refers to the effects of an economic activity that involves costs and benefits to people not involved in the decision about whether to have the activity; thus, the effects do not get registered in the price system and are not properly allowed for in private decision making.

An obvious external cost is environmental damage. If burning some kinds of coal generates pollution that adversely affects citizens other than those burning the coal, and if the people making the decision to burn coal do not have to pay for these damages, they will be making their decisions on the basis of imperfect information and will burn the wrong kinds of coal. They will also think that coal is cheaper than it really is.

It is important to notice in this example that, strictly speaking, the problem is not that coal generates environmental damage, nor even that coal companies are greedy, but that the legal system does not result in information about environmental damage being reflected in prices. Polluters are not required to pay for pollution, and until recently they were not required to clean it up.

If we knew that the environmental damage generated by burning high-sulfur coal was trivial, then it would be inefficient to incur massive shutdowns of the coal industry and extensive new investments in other expensive forms of energy to correct a trivial pollution problem. On the other hand, if we knew that the atmospheric sulfur generated by coal burning caused very serious and widespread damage, it would be worthwhile to incur heavy costs to avoid it.

The real defect involved is that the price system allows the coal burner to take account of only the costs he pays, such as the fuel costs. If he also had to recognize as a cost the damage inflicted on others, his normal profit seeking would cause him to make little allowance for it if it were trivial or massive allowance if it were serious.

Externalities can involve benefits as well as costs. It may be that investment in the domestic industries producing energy resources provides a net benefit to the United States by improving our long-run national security. An investor in the oil business cannot directly charge anyone for this benefit; consequently, there would be underinvestment in oil.

Another possible set of defects in a price system may be the peculiar circumstances that arise when a product is manufactured from an exhaustible resource. When a product is made by machinery, a producer has a pretty good idea of what it costs to use up the machine, namely, the price of buying another one. For products such as oil and gas there is more uncertainty and risk involved in whether we will find deposits to replace the ones being used up. How does the market assign a price for resources and will it be the right price? That is, will it lead to resources being used up too rapidly or too slowly?

It is significant that for a long time Congress has apparently been convinced that for exhaustible resources there were defects in the pricing system that required government intervention in the form of tax incentives for minerals.

Other important defects in a price system arise when there is a lack of effective competition. It may well be that there is imperfect competition in the energy industries.[5] The whole matter of competition and monopoly cannot be the major focus of the present work, because taxes and subsidies are a very inefficient instrument for dealing with monopoly problems.

The concern of this work is with the way in which the economy responds to price and income signals. As necessary, we will talk about the efficiency or the defects of the system in terms of the existing structure of competition and/or monopoly. This is clearly an important part of the picture in, for example, the international oil market, where the cartel formed by OPEC is the critical factor in explaining price and output developments.

Whether domestic or international markets would work better if they were more competitive and how they could be made more competitive are intriguing questions, but we must leave them to others.

The "defect" of the price system that has drawn the most attention in public discussions in 1974 is the possibility that large price changes in the short run can create windfall profits for sellers. The same high prices for such important goods as fuel and electricity are an unbearable burden on poor people. When a shortage emerges fairly suddenly, say, as a result of a war in the Middle East, the market adjustments are harsh indeed, which is why many people argue that in such a circumstance the government would have to control price and ration or allocate supplies and/or impose a windfall-profits tax. Long-term price and profit controls and allocations will aggravate rather than cure the basic shortage problem. Part of the energy problem, then, is that the normal market responses through higher prices will create harsh income redistributions.

We are not here trying to list all of the defects in the energy price system. Rather, we are only indicating the kinds of problems that might lead to a recommendation for government intervention in the market, and suggesting a way of thinking about government intervention, namely, how it deals with the defects of the price system.

1.4 TAXES AND SUBSIDIES AS CORRECTIVES FOR PRICE DEFECTS

Broadly, the government pursues various kinds of policies with regard to the private price system. In the energy area our complex system of laws establishes particular ownership rights to subsurface minerals and brings about different mineral prices and outputs than would have occurred if the ownership rights had been defined in another way. The The structure of price regulation established for privately owned public utilities is another case in which fundamental government rule-making establishes an environment that leads to a particular set of prices.

We cannot treat all energy issues in one book. In talking about taxes and subsidies we are largely putting aside, as already given, government policies concerned with the structure of the energy industry, the basic ownership rules, and the price-making institutions. Like the problems of competition and monopoly, these are important matters, but they must be examined separately.

The important aspect of a tax or a subsidy is that it changes a price and, thus, an income that would otherwise be established in the economy. The excise tax on gasoline for the highway trust fund is an obvious increase in the pump price of gasoline. Less direct tax measures, such as special income tax allowances for oil production, also act as a price change. They tell the producer of oil that the net income after tax will be higher than it would have been with existing prices and regular taxes. Such policies as excise taxes on gasoline and special income tax allowances cause prices to change; they also cause producers to act differently in response to given prices, which in turn can change outputs, prices, and costs. Frequently, it is a difficult matter to say what these differences will be.

This way of looking at tax and subsidy policy might appear strange if people think of tax policy as concerned solely with raising revenue to pay for public services. Actually, contemporary public-finance economists describe the objective of a tax system not as raising revenue but as reducing private purchasing power to free resources to meet government expenditure demands.[6] This description brings home the point that taxes necessarily interfere with prices and market decisions; the issue is merely which decision will be interfered with.

Recognition that taxes must interfere with the market suggests a different approach to tax policy. Our general argument is that taxes should interfere with market decisions as little as possible, that is, they should be even-handed or neutral, unless there are defects in the market price system. If there are such defects, an unequal tax,—a tax preference—or a subsidy should be evaluated as a potential government policy to offset them.

In taking this approach, we reject the inflexible assumption that taxes should be neutral or that the income tax must be comprehensive. Rather, we look at the evidence of defects in the market price system and examine, on the basis of theoretical and statistical analysis, what certain unneutral tax provisions can be expected to accomplish and what they actually do accomplish.

While this approach clearly admits to a willingness to consider tax or subsidy provisions that tinker with economic Mother Nature, the fact that we have identified the tinkering is a very important constraint.

We must now examine the market processes that brought about the present energy industry and ask certain questions: Why will the market not reach the right decision if left to itself? How effective is the proposed policy in dealing with the market defect? Are there better policies available?

Another basic consideration that needs to be noted is that in popular discussion about taxes much attention is given to who pays more or less than his "fair" share, and little attention is given to the subtlety of price intervention. The question of equity cannot be addressed separately from the analysis of taxes as effecting price changes and, in turn, business decisions.

To cite one example: It is commonly asserted that oil companies pay U.S. income taxes at a lower rate than other companies. Our concern is to go beyond this kind of "fact" to ask what happens as a result and whether these tax differences produce commensurate public benefits.

To look at it another way, the government might be thought of as spending money, through tax concessions, to get some public benefit, for example, a more secure domestic oil supply. The question of how fair the tax concessions are should be viewed in relation to what the government gets for its money.

1.5 A NOTE ABOUT THE EMPHASIS OF ECONOMISTS ON TRADE-OFFS

Economists who work with public policy issues have a professional bias, which makes them like everybody else. We are suspicious of wonderful solutions that help everybody and hurt nobody. When we struggle with public policy issues, we talk about trade-offs. To get more of one thing, you usually have to accept less of something else.

There are two reasons why it is important for readers to understand this emphasis on trade-offs. One common idea in popular discussion of public policy is the notion that the economy "needs" such and such. The layman usually makes his point by citing the average yearly increase in per capita consumption of energy (or water or invested capital) in some past period. Then he extends the rate of growth for, say, 17 years, multiplies it by the population expected in 1990 and produces a figure for "needs" of energy in 1990 (or needs for water or invested capital).

From the economists' viewpoint this is nonsense. We do not *need* such and such amount of energy in 1990. We are faced with trade-offs. To get this much energy we will have to accept less of something else. There may be some evidence that suggests that a certain amount of energy is a good buy. In fact, it may be such a good buy that we would be better off with even more energy. The important question is not whether energy is good or whether it is needed but what the trade-offs for more or less energy are.

It is irrelevant to say that our system would die without an energy industry and that, therefore, energy is a need. No one is even talking about abolishing the energy industry. We are talking about having more or less energy. This is not actually a need; it is a choice, and choice is related to the price of energy.

To illustrate this point: It can be observed that one major energy consumer is the automobile. Most European countries tax gasoline far more than enough to cover highway expenditures, to the point that the pump price is two to three times as high as it is in the United States. Gasoline consumption in these countries is about half as high in relation to income as it is in the United States. We can see, then, that higher energy prices lead people to consume less energy.

Choice is also inherent on the supply side. It is usually a fruitless course of action to wait for the perfect energy source. Technology may yet find one source that is plentiful, cheap, safe, and nonpolluting, but until this miracle of research occurs, we must plod ahead in human fashion to make do with what we have. That is, we must seek efficient marginal trade-offs among various less-than-perfect sources.

1.6 EXISTING TAX AND SUBSIDY POLICY

In the practical world policy analysis must start with an existing set of government policies. These are described in more detail in Chapter Two, but it will be useful to summarize them briefly here to make our summary of conclusions more comprehensible.

The significant tax policies related to energy are concessions that reduce the levy on income earned from producing natural resources. The most important of these benefits is percentage depletion, which provides for tax exemption on a portion of net income received from producing a natural resource. The natural resources that are most valuable in the energy industry are oil, gas, and uranium, which qualify for a 22 percent depletion rate; oil shale, which qualifies for a 15 percent rate; and coal, which has a 10 percent rate. Producers of natural resources also get another income tax benefit; special rules allow for current deduction of some investment expenses related to exploration and development.

Another income tax differential important in the energy industry is the tax exemption of publicly owned power companies. Companies owned by state and municipal governments also enjoy the benefits of lower interest rates because they can borrow by issuing tax-exempt municipal bonds.

Privately owned public utilities in the energy industry receive a lower rate of investment credit on the purchase of machinery and equipment than business generally, but they benefit to some extent from a complex tax policy designed to induce regulatory agencies to handle excess depreciation in a way that is favorable to the companies.

Oil and gas companies are heavily engaged in foreign operations. Typically, the foreign governments impose very heavy taxes on U.S. companies, although these taxes are in some respects different from ordinary income taxes. Under U.S. tax law, those oil and gas levies are, nevertheless, treated as income taxes and can be subtracted from the U.S. tax due on the same income.

In the United States a substantial excise tax, both federal and state, is imposed on the sale of gasoline; most of this is closely related to expenditures on highways. The total state and local tax on the purchase of energy is somewhat heavier than that on other purchases, which are covered under general sales taxes. Typically, local governments tax the property of privately owned public utilities more heavily than they tax property in general. Some states also tax the energy industry more heavily by imposing a severance tax on energy resources taken from the ground.

The federal government also directly subsidizes the energy industry, most conspicuously by expenditures on energy research, which now come to about $750,000,000 in fiscal year 1974 and are programmed to rise substantially in 1975.

1.7 SUMMARY OF CONCLUSIONS

This summary is keyed to major policy issues of taxes and subsidies related to energy. Approaching the task this way will slight the contents of Chapters Three and Four which are groundwork for the specific policy analysis, but the thrust of those chapters comes through in the final conclusions. The remaining questions addressed in the summary parallel Chapters Five through Twelve.

Should we provide tax or other incentives for development of domestic energy resources?

The tax benefits of percentage depletion and deduction of intangible drilling expenses on successful wells, which principally aid the oil and gas industries, should be terminated.

Domestic oil and gas are in short supply. In recent months, with general price controls, we have experienced shortages; without the controls, we will experience higher prices. Tax incentives to producers of oil and gas meet the situation by increasing the supply, which, in turn, keeps the price lower than it would have been.

This is a bad policy for several reasons. In the first place, to the extent that it increases supply with less price increase, the policy encourages consumers to use more oil and gas than they would if the price were higher. Producer incentives earned in the marketplace through higher prices that lower consumption are more sensible than producer incentives through lower taxes, which entail lower prices and more consumption.

A more specific irrationality of the policy of "encouraging" oil and gas consumption, that is, reducing the price, is that in the long run there are alternative sources of energy—uranium, liquefied or gasified coal, kerogen from oil shale, and sunlight, to mention the main ones. Our present tax structure provides a very substantial incentive for energy produced from oil and gas because these resources are scarce and have a high value in the ground; it provides a very small incentive for energy from cheap resources, where the energy is obtained by applying manufacturing processes. In the extreme, if a firm could take common dirt and manufacture it into the equivalent of a scarce resource, the present tax incentives of percentage depletion and deduction of intangibles would still favor the producer (or the use) of the scarce resource over our hypothetical firm that achieves its product by manufacture. The tax provisions serve to lower the price on the scarce resource and to increase the royalites of the landowner. The development of substitute sources is retarded.

We conclude that percentage depletion and deductibility of intangible drilling expenses have increased the output and lowered the price of petroleum products and of natural gas, but they have done so by maintaining artificially low prices that discourage development of new fuels from other resources.

Repeal of these provisions would increase oil and gas prices and reduce payments to landowners. The consuming public would gain more from an increase in federal tax receipts and the opportunity to reduce other taxes than it would lose in higher prices for fuels. In the long run, the public would gain from the increased productivity due to a more rational allocation of resources between competitive energy modes.

A government program that reduced prices of natural resources would be defensible if a careful analysis of the price mechanisms that direct investment in energy resource areas revealed reasons to expect that free market prices would tend to produce under investment in natural resources.

There are three major problems with free market energy prices that must be faced:

(1) A free market is likely to result in underinvestment in research on new technologies, such as coal gasification, extraction of oil from shale, and the development of solar energy transfers. The government recognizes this problem in principle by spending federal funds on research. Research subsidy arrangements should make provision for some governmental participation in any future windfalls that might be generated by successful research.

(2) Drillers have limited information about location of oil and gas deposits, and drilling produces valuable information that is frequently exploited by people other than the driller. A variety of policy changes are appropriate to deal with this, including (a) the leasing of larger blocks of property when federal lands are put out for bidding, (b) a subsidy for information-seeking activities of oil- and gas-well drillers (geological and geophysical studies) with the information generated to be made available to the public, and (c) possibly a subsidy for

exploratory oil and gas wells defined in a restrictive way, for example, wells located at least a mile distant from a commercial well would be eligible.

(3) Present market pricing institutions do not provide reserves of energy resources against possible interruptions of foreign sources: the security problem. A security problem arises when foreign oil is cheaper than U.S. oil—the case prior to 1973. Then we had a quota on oil imports, which protected the U.S. price, and large tax benefits for oil producers, which kept domestic prices somewhat lower. When the foreign price goes above the U.S. price to the point of creating what President Nixon has called windfall profits for U.S. producers, tax relief is not needed. It is possible that in the future foreign oil prices will again be low (if, say, the OPEC cartel breaks up). If this happens, we think that there are two kinds of policies that will deal with the problem efficiently, and that they should be used together. One is to maintain a strategic inventory of crude oil reserves. The other is to impose a special tariff on oil imports. Unlike domestic subsidies, such as percentage depletion, these policies do not increase consumption rates—although the import tax, by raising the price of imported oil, extends an implicit subsidy to all domestic energy resources. The tax on oil imports is an ideal source of revenue for meeting the expenses of an inventory program, because it puts this cost on oil users whose interests are being protected by reducing the chances of an interruption of supply.

Should we provide tax or other incentives for development of foreign energy resources?

Tax treatment of U.S. companies in the oil business abroad must be considered in relation to general treatment of foreign investment income under U.S. tax law. Our law accepts the principle that the foreign country in which income is earned should be the first to impose income tax and that the amount of the tax can be subtracted from tax due to the United States on that foreign income.[7] (The foreign tax cannot be subtracted from taxes due on domestic income.)

Regarding foreign investments in oil and gas, there are three problems arising out of the very heavy income taxes imposed on oil production by member countries of OPEC. These taxes are so high that there would be no U.S. tax even if percentage depletion on foreign oil were denied.

One specific issue is that under present law excess taxes paid to the OPEC countries can give rise to foreign tax credits that reduce tax payable to the United States on activities, such as shipping, that incur relatively low foreign taxes. This misuse of the foreign tax credit should be terminated.

Another issue relates to the treatment of start-up losses. When a U.S. oil company begins operations in a new country, it will show net losses for several years because of the generous deduction of drilling expenses. These losses may be deducted from income taxable in the United States; thus, the losses are compensated to the extent of 48 percent by the U.S. Treasury. When the operation in this new country becomes profitable, the company does not have to pay a tax on that profit to the U.S. Treasury because of accumulation of the foreign tax credit. This one-way drain on the U.S. Treasury should be terminated by denying the foreign tax credit until an amount comparable to prior loss deductions is taxed under U.S. law.

The final issue relates to the tax treatment of the basic profits on production in the OPEC countries. Although the application of the foreign tax credit would make these profits nontaxable in the United States, the fact is that the OPEC charge, while called an income tax, is not in its basic structure an income tax, but an excise tax. In the short run, we do not see any strong advantage to the U.S. in changing this treatment because the oil companies of other nationalities are also protected from home-country taxes; imposing a significantly higher tax only on U.S. companies would damage our interests. In the long run, however, we do see advantages in pursuing discussions with the home countries of other producer companies, looking toward a coordinated tax policy.

How should the tax law apply to public utilities?

A major change called for is the elimination of the present favorable tax benefits for publicly owned utilities (income tax exemption plus access to tax-exempt interest financing). Given the long-run prospective of price increases related to energy shortages and of environmental problems related to energy, it is anomalous to provide what is in effect a subsidy for electricity consumption through public power.

There are, however, some institutional problems associated with applying federal income tax directly to instrumentalities of state and local governments. For one thing, these operations could avoid tax by pricing closer to cost, which aggravates the excessive use of energy. An alternative to direct tax would be an excise tax equivalent to an income tax. We should also extend the concept of taxable industrial development bonds (Section 103(c) of the Internal Revenue Code) to public utility bond issues, thereby depriving public power of its access to the artificially low interest rates on state and local tax-exempt obligations.

Another issue relates to the tax law applicable to private utility companies in electricity and gas distribution. The law allows these companies only half of the investment credit, but the full benefit of accelerated depreçiation. The law does, however, go to some pains to require the commissions that regulate public utility prices to handle these tax incentives by a process called "normalization," which gives short-run benefits to the companies, rather than by "flow-through," which gives short-run benefits to the customers.

Reviewing these policies, we find that there is justification for allowing little or no investment credit for regulated public utilities because the allowance tends to push the companies to over-invest in capital equipment, which is already excessive. Also, in the long run utility companies would benefit from flow-through of tax benefits, and consumers would probably not lose. This suggests that the ostensible purpose of requiring regulatory commissions to allow normalization, that is, to encourage investment by public utility companies, is not being achieved. In the long run, investment would be higher under flow-through. From the narrow objective of dealing with the energy problem, the policy of requiring normalization seems attractive. It keeps energy prices up and holds down electricity production.

How should tax and subsidy policy deal with pollution?

In the face of violent disagreements about the war in Vietnam in the late 1960s, people seemed delighted to agree on cleaning up the environment. There was general agreement on having "tough" laws strictly enforced before there was much scientific agreement on the technology of the pollution problems themselves.

We are beginning to recognize that environmental goals are expensive and can seriously conflict with other goals, particularly energy goals. This does not mean that one or the other must give way completely, but rather that we must have institutions for making intelligent decisions in both areas. We need to find the best pollution control techniques, essentially choosing between taxes and direct regulation.

Upon close analysis it turns out that pollution taxes and direct regulation of pollution have much in common. To clarify the relation between the two, it is necessary to refer abstractly to extreme forms of the two policies. A pure tax approach would penalize all pollution of a certain character, say, sulfur oxides [SOx].[8] In a pure regulation system, the government would assign a permissible level of pollution, then attach prohibitive fines to violations.[9]

A disadvantage of the pure regulation approach is its all-or-nothing character. The implicit posture of regulation is that a firm that violates will face prohibitive penalties, which puts great weight on the decision about the critical point. In 1973 regulators decided that they had selected the wrong pollution-control requirements for automobile design; despite a poor performance no one was blamed and the regulations were changed. In other situations litigation over one aspect or another of the critical requirement can delay its application for years. These cases suggest the merit of the milder tax approach, which makes pollution expensive and more pollution more expensive. A pollution tax exerts a slow, steady pressure on polluters that will produce results over time; more spectacular efforts, that is, to impose prohibitive fines at critical points, often go awry.

Combinations of tax and regulatory control are feasible and promising. In cases where we know that potential damage to the public increases more than proportionately with pollution levels, there is real substance in the regulatory concept of drawing a line and saying, "If you cross this line, we will fine you heavily." It is in cases where the marginal damage from a little more pollution does not increase much that it is very hard to rely fully on the critical-point concept of regulation. In these cases it is very promising to put reliance on the slow, steady pressure of a tax.

The real advantage of the pollution tax is that it permits, more effectively than regulation, adjustments "at the margin." Pollution reduction has a value to society, just as more energy has a value to society. What we want to achieve is a level of pollution reduction that has a value greater than its cost. A pollution tax represents a social concensus of the marginal value of pollution reduction. The tax tells industry that the cheapest pollution reduction should be undertaken. The tax may be very high if the public attaches great value to the pollution reduction.

The notion of seeking marginal adjustment should be extended to the question of setting regional pollution policies. In our view a good regional policy would use a technique that increases pollution control until the marginal cost of further pollution reduction equals the marginal benefit of reduction. Under such tests the pattern of efficient results should be in the direction of, for example, requiring highly sophisticated stack-gas cleaners for sulfur in a metropolitan area, where lots of people breathe the same air, and the use of low-sulfur coal in sparsely populated areas.

We think that the existing pollution control program involves an inefficient regional policy, one that involves excessive energy costs for the environmental benefits achieved. There are, for instance, excessively strict ambient air standards in lightly populated regions and excessive concern with the concept of "degradation" in regions with very clean air.

Should we use selective excise taxes or rationing to modify energy consumption patterns?

By and large, a general tax to reduce energy demand is not a promising way of dealing with an energy shortage, for the same reasons that tax incentives for producers are not efficient. Shortage causes either rising prices and/or some kind of rationing—if only of the first-come-first-serve variety—to reduce the quantity demanded by consumers. If the market price is permitted to rise and suppliers are permitted to enjoy higher prices, then we would expect suppliers' profits to rise and an incentive to expand output (very likely reversing some of the short-run price rise). If a consumption tax is imposed, it effectively raises price. It can reduce the level of demand enough to avoid rationing. The tax-related price increase does not increase supplier profits; thus, it generates no incentive for increasing supply. The rationing approach has the same defect. It reduces price and cuts demand. The demand cut is arbitrary, and the absence of a price increase serves to hold back the growth of supply.

Our critique of a general energy excise tax does not extend to taxes on particular forms of energy consumption when it can be shown that market circumstances lead to over-consumption. In this case we have a market that should not be provided for so generously. Taxes on automobile driving are a case in point. Several circumstances lead us to believe that the automobile imposes net costs on people other than the driver. Drivers, then, see only part of the cost and over-consume. The problem could be remedied by a good highway toll system or a parking tax. An increased gasoline tax would be a third-best solution.

It would also be sensible to resort to taxes on particular energy sources when demand rises so rapidly that serious windfalls are generated. Thus, in the matter of deregulating natural gas, a fairly high tax rate applied to well-head prices in excess of, say, 50 cents per thousand cubic feet (mcf) would limit windfalls without interfering with the basic price function of bringing about greater production of natural gas.

This proposal has a broad similarity to the windfall-profits tax proposed by President Nixon for crude oil in December 1973. The President's proposal would tax oil on the well-head price, less $4.75 per barrel, at rates that progress with the price. The difference between the Administration's proposal and our proposal is that we suggest that the tax base be measured from a relatively higher price, say, the long-run supply price. In the explanation of the President's proposal provided by the Treasury Department, the long-run supply price is estimated at $7 per barrel, and so the tax would reduce production from wells that would have a marginal cost of production over $4.75. More significantly, our proposal differs from the President's version since ours would be combined with repeal of the special tax advantages extended to oil and gas under present income tax law.

What about the income distribution effects of energy policies?

Allowing gasoline, fuel oil, natural gas, or electricity prices to rise will, taken by itself, hurt poor people. This circumstance will be used by some as an excuse for such policies as tax incentives for energy resources, on the specious argument that these policies will help the poor.

We must keep in mind that government has ample means at its disposal to help poor people directly. It can increase welfare payments, lower payroll taxes or income taxes on the poor, provide more food stamps or more low-rent housing subsidies. Ultimately, the poor can be hurt, however, by a lot of economic programs that involve much social waste but are promoted on the grounds that some of their incidental effects seem to help poor people.

Because it is a feature of our political process that programs are voted separately, there are problems in getting the public to recognize the connection between decisions on energy policy and programs on welfare. It may be useful to state the connection symbolically so that an income distribution measure can be identified as a component of "the energy program."

Such a feature is available by an extension of the tax refund idea suggested in Chapter Ten. We could provide an energy allowance that would give every family a cash payment equivalent to the expected increased costs of energy at, say, a moderate income level. The cost increases specifically identified would be the burden of any increased energy taxes, of increased heating costs due to deregulation of natural gas prices, of increased oil and gasoline prices.

An allowance of this sort could be claimed by most families as an income tax credit and could be phased out at higher income levels. Most families file a tax return if only to claim withholding refunds. The energy allowance would be refundable in cash if there were no tax liability. Welfare and social security offices could distribute simple forms for people who do not use income tax returns to assure that they would get their credits.

A further word needs to be said about the income distribution effects of changes in producer incentives. At present, because of rising world oil prices, repeal of percentage depletion and the intangible deduction allowances would not cause drastic losses to oil companies. Repeal of percentage depletion would reduce the after-tax income of present recipients of royalties on oil and gas lands. In many cases these royalty incomes are mere windfalls that come unexpectedly as a by-product of land ownership. In other cases the land was purchased in anticipation of tax benefits. A change in tax treatment creates some unfair losses in the latter instance. Congress did not, however, provide special relief after the reduction in percentage depletion in 1969, though some temporary relief for pre-existing royalty contracts seems to be warranted.

As for new royalty contracts, we should take into account that in a free U.S. market there would be dramatic price increases in both oil and gas in the short run and possibly in coal and/or shale in the long run. In the oil industry the increases could result from world price increases; in gas from deregulation; in coal and shale from new technology. We think it is important that prices be allowed to respond freely to market developments.

A political barrier to reliance on a free market will be allegations that certain groups, especially landowners, reap large windfalls. We think that a means of dealing with windfall incomes that is superior to price control is to impose a federal severance tax in cases where royalties appear to be rising excessively.

Should there be an energy trust fund?

The issues raised by proposals for an energy trust fund are complex and get down to the basic questions of how efficiently expenditure decisions are made under our existing budget institutions. In this study we have not undertaken a detailed enough cost-benefit analysis to warrant strong positions on particular expenditure decisions.

Even if we assume that benefits would be strong in relation to costs, we think that programs for research, for stockpiling, and for urban mass transit will do badly in competition with the established budget programs. The budget conflict is likely to be aggravated by a peculiar feature of two of the large expenditure components in the regular budget—defense and welfare. In defense and welfare issues there are people who advocate more spending and people who feel particularly threatened by spending levels they consider "too high." In the case of defense the concern is that more defense means greater threat of war; in the case of welfare, that more welfare undermines the work ethic.

We think that this combination of circumstances is related to the difficulty of getting general tax rate increases. We would characterize the two-year delay in getting a tax increase to pay for the Vietnam war as the combined effect of attitudes of the opponents of the war and the opponents of the Johnson welfare programs.

An energy trust fund, that is, a fund earmarked for special

use, financed by, say, a general tax on energy would create a political tax rate issue largely separate from the normal budget battles over defense and welfare spending. The advantage of the trust fund is not that an energy tax is a good user tax; it is not. The advantage is that it creates a separate political issue, one that would probably be resisted less than a general income tax increase.

We also endorse user taxes in situations where the tax operates closely enough to a price to provide useful information on what should be spent on a particular category. Most of these possibilities appear to arise in the pollution area, where the tax might reflect how much should be paid to compensate sufferers.

A particularly useful earmarking device would relate the proceeds of a tariff on imported oil to the costs of a program to stockpile crude oil. This system would key the cost of stockpiling to those for whose benefit the stockpiling is undertaken—oil users. Also useful would be to earmark increased automobile user taxes to expenditures for mass transit.

What should we do about imports?

Fortunately, oil import quotas are dead, but the concern about oil import dependence persists. This concern rests ultimately on the possibility of a serious interruption of supply due to hostile interference with shipping of the type that occurred during World War II, or due to concerted action by several OPEC countries as retaliation for, say, support of Israel. This concern, as we have already described, points to a domestic resource strategy of imposing a tariff on imported oil and using the proceeds to stockpile crude oil.

Another dimension of the import problem is what strategy to use in the face of the OPEC cartel. In addition to political negotiations to relieve embargos, the basic enonomic strategy of the U.S. *vis a vis* OPEC should entail two kinds of efforts:

(1) To do things that cause the ratio of future to present crude oil prices to be lower;

(2) To maintain a long-run "elastic demand curve" for oil imports. The expectation of lower future oil price—due to rapid development of nuclear or solar energy, for example—would cause the owners of oil reserves to produce more now while prices are high and save less for later when prices are expected to be lower. Maintaining an elastic demand curve for imported oil means that producers will have more to gain from a low price-large output strategy.

The general thrust of these two efforts is somewhat contradictory. By making a determined effort to achieve domestic self-sufficiency in oil, we are likely to convince OPEC that it is in their interest to sell us oil at a lower price so that we do not become completely independent of their oil. A lower oil price strategy will only work for OPEC if, in fact, their customers will buy more oil, that is, if they do not revert to oil import quotas to protect their new industries producing substitutes for OPEC oil.

Wisdom will be called for. A balance must be struck between reasonable protection for expanding industries in the face of the risk that OPEC oil may become cheap again. The degree of protection must be moderate to avoid saddling Americans with very heavy energy cost. To the extent possible, we should be willing to abort particular programs for developing high-cost energy substitutes and incur the costs of compensating labor and capital for their sunk committed costs. Assurance that these costs will be covered, if the need arises, will hasten the development of substitutes.

If we can deal with the short-term competitive impact on U.S. industry of a return to low cost foreign oil, we should not be concerned with the notion popular in the early 1970's that prospective oil imports will involve a disastrous balance of payments deficit for the U.S. The reason is that we are confident that the events of the last few years have put us well on the road to much more flexible exchange rates and that exchange-rate adjustment does assure adjustment of balance of payments deficits.

There is an economic problem in a heavy increase in the U.S. balance of payments deficit on oil, even if the problem is not a deficit that will bankrupt the U.S.: the net cost to the U.S. of massive oil imports is more than the amount paid for the oil. The net cost includes the high cost of imports, and the lower proceeds from exports due to whatever shift in exchange rates can be attributed to the oil imports themselves. This suggests that it would be in the interest of the U.S. to impose an additional tariff on imported oil. Coupled with an inventory reserve, a tariff-reserve program would give us some defense against future cartel price manipulations and an assurance of a steadier world oil market.

NOTES TO CHAPTER ONE

[1]For some perspective on these matters, see Marc Roberts, "Is There an Energy Crisis?" *Public Interest,* Spring 1973.

[2]For the reader who is eager to get ahead with the story, much attention is given later to whether the existing price system is ideal and what is needed to improve it. It is useful first to talk very generally about the role of prices.

[3]For a useful historical perspective on how the United States market has adjusted to changing prices of different energy forms, see Nathan Rosenberg, "Innovative

Responses to Materials Shortages," *Proceedings of the American Economics Association,* May 1973, pp. 111–118.

[4]We will try to avoid economic jargon except where a technical word very efficiently captures an important idea. Externality is an important concept for human affairs, and the word should be in common use. Love, for example, could be defined as acting so as to provide external benefits.

[5]Monopoly is recognized and regulated in the electricity and gas industries. A staff report from the Federal Trade Commission contended that lack of competition in the oil industry, which is partly related to earlier government policy, is responsible for some of the current shortage problems and is a matter that must be dealt with to assure efficient future adaptation of the industry to meet consumer demand. *Preliminary Federal Trade Commission Staff Report on Its Investigation of the Petroleum Industry,* Committee on Interior and Insular Affairs, U.S. Senate, Serial No. 93–15 (92–50). Washington, D.C., Government Printing Office, 1973. The energy monopoly issues are also dealt with in a study in the Energy Policy Project series by Thomas Descheneau, "Competition in the Energy Industry."

[6]Carl Shoup, *Public Finance,* Chicago, Aldine Publishing Co., 1969, p. 61.

[7]Subtraction of foreign taxes from U.S. taxes on the foreign income is called the foreign tax credit. It is generally accepted as reasonable by tax writers, although it has been criticized as excessively generous. See Peggy Musgrave, *U.S. Taxation of Foreign Investment Income: Issues and Arguments,* Cambridge, Mass., Harvard Law School, 1969. Chapters 5 to 7.

[8]That is, a sulfur tax would apply to all sulfur emission into the air. [9]This description should, but will not, put to rest the claim that taxes are "a license to pollute." The pure tax approach allows no "free" pollution; the regulation approach does.

Chapter Two

Present Taxes and Subsidies Affecting the Energy Industry

2.1 DIFFERENTIALS

The total of federal, state, and local taxes paid by the energy industry is not a particularly relevant number. What is important from the standpoint of energy policy is how existing taxes and subsidies fall differentially on the energy industry. It is these differences that can distort energy prices and incomes and thereby divert energy outputs.

This chapter describes the substantial tax and subsidy differentials that affected the energy industry in 1973. In sections 2.2 and 2.3 there is a discussion of the relationship of efficiency and fairness in tax policy plus some judgment on how the major differentials change. In section 2.4 there is an estimate of the price and output effects of the oil benefits.

Income Tax of Natural Resource Producers

The best-known tax provision affecting the energy industry is percentage depletion. Under the income tax provision other industries are permitted to deduct the cost of capital assets that are used up in production (usually called depreciation). Natural resource producers, however, have the privilege of treating a part of their income as tax-exempt, through percentage depletion, even if it far exceeds the cost of the assets (the wells or mines) being used up.

Percentage depletion is important for oil and gas. These products have the highest percentage depletion rate—22 percent of gross income. Percentage depletion is based on the value of resources as they come out of the ground, and only to a minor extent applies to the value added by processing and transportation.[1] The percentage depletion rate for uranium is also 22 percent. The rate on oil shale is 15 percent; on coal it is 10 percent.

Percentage depletion has some limitations that make it less generous than it sounds. In the first place, the allowance can only be applied to 50 percent of the net income from the property.[2] Further, a taxpayer using percentage depletion gives up the opportunity to deduct cost depletion on some relatively small part of the actual costs that were incurred in developing the mineral property. Finally, percentage depletion is slightly reduced in value for many taxpayers because, since 1969, those who use it may be subject to a minimum tax.

Since our interest is in the policy, not the mechanics, it will be more useful to forget the statutory rate of percentage depletion and think of it as equivalent to a net tax benefit after the limitations. For oil and gas the net benefit overall is about 15 percent of the gross income from producing the natural resource; for uranium it is about 10 percent; for coal, about 5 percent.[3]

These figures relate only to income from production of the crude raw material. In the case of oil and gas the value of crude is relatively high as it comes from the ground compared with its value when it is finally converted to energy. For uranium and coal the production value of the crude material is not as high in relation to final value. Some estimates of the importance of this difference are offered below.

A further income tax differential arises from special rules that allow current deductions for intangible drilling and development costs for oil and gas and certain exploration and development costs for other minerals.[4] The differential benefit occurs because capital investments in other businesses are deductible only gradually, as the investment asset wastes away. A current deduction is worth more than a deduction spread over the life of the asset because of the "time value" of money. As a rule of thumb the current deduction is twice as valuable.[5]

The current deduction treatment is even more favorable to the industrial taxpayer than a mere speed-up of deductions because percentage depletion itself is a substitute for the gradual charge-off of wasting assets. Using the combination, companies can deduct many of their capital costs as intangible drilling expenses or development expenses as they are incurred and also use percentage depletion as a substitute for cost depletion.

At this point, the detailed mechanics are less important than a summary of just how much the tax benefits amount to. One way to look at these benefits as they apply to different fuels is to describe them relative to the investment cost. Another way is to describe them as a percentage of the price of energy from that fuel.

The two ways of describing the tax benefits, in a practical sense, correspond to the two major effects of the tax provisions: an increase of investment in a particular resource (drilling more oil and gas wells or developing more coal mines, for example) and a reduction in the price of the particular mineral product.[6]

To the extent that natural-resource tax provisions provide an incentive to invest in natural resources it would appear useful to describe the provisions in terms of an equivalent investment credit, which could then be compared directly with the investment credit applicable to business investment generally.

The special tax benefits of percentage depletion, deduction of intangibles for oil and gas, and the normal investment credit for tangible drilling costs are equivalent to an investment credit of 49 percent.[7] An earlier analysis of the tax benefits for investment in mineral production also expressed the outcomes in relation to capital investment and concluded that these benefits were in the general neighborhood of a 50 percent investment credit for oil, gas, and coal.[8]

The ultimate concern of energy policy is not with investment in energy resources but with the sale and use of energy. The second way of looking at how natural-resource tax provisions apply to different fuels, therefore, is to describe them relative to the final price of energy obtained from that resource. It is particularly relevant to the extent that the benefits are passed along to consumers in price reductions.

When we look at natural-resource tax provisions this way, we can see conspicuous differences among fuels. The differences arise in part because percentage depletion is based primarily on the value of resources as they come out of the ground; in general, it does not apply to the value added by refining, transportation, wholesaling, and retailing. As far as the tax benefits of expensing drilling and development are concerned, capital investment involved in development differs greatly among resources. In general, oil and gas production entail a relatively high share of capital cost compared with labor cost; production of coal, uranium and oil shale entail a relatively low capital cost compared with labor cost.

Let us consider the example of a major fuel user—steam electric plants—and ask the question, "If all the savings from the natural-resource tax provisions were passed along to the consumer in the form of price reduction, what would be the percentage reduction in fuel cost for electricity generated from each fuel?"[9] Some answers are:[10]

oil	13.2%
natural gas	11.5%

coal	3.4%
gas from coal	1.4%
uranium	2.8%
oil from shale	4.5%

We think that these differences in delivered prices as related to the tax benefits are much more important for energy policy than the apparent uniformity of the benefits in terms of invested capital. To see why this is so, consider that capital costs are 10 percent of the value of mineral production for coal and gas.[11] Even if the natural-resource tax provisions cut the capital cost in half for both industries, the change is bound to be more important in the oil and gas industry, where a far greater cost would be affected.[12]

It is clear, then, that a fatal flaw of the present income tax provisions as energy incentives is that they work very unevenly among the different resources that go into energy production.

Income Taxes and Public Utility Companies

Another major income tax differential in the energy industry that we must deal with is the income tax exemption of publicly owned power companies. Most of these companies, which are owned by state and municipal governments, also enjoy the benefits of the lower interest rate on tax-exempt municipal bonds. These two differentials can lower the cost of electricity to the consumers of public power by one-fifth. The treatment of publicly owned companies in regard to property taxes is hard to pin down because there is considerable provision for payments in lieu of property taxes.

For private companies involved in the generation and distribution of electricity and the distribution of local gas, the income tax law is about neutral. Though private companies face the disadvantage of having only half the rate of investment tax credit of business generally, there is some offsetting advantage in complex tax provisions that induce regulatory commissions to take account of investment tax incentives in a way that is favorable to the companies.

Income Tax on Foreign Operations

The treatment of foreign taxes under United States income tax law is important for the energy industry only in the case of overseas investment in oil and gas. We limit our comments in this area by making the initial assumption that the general technique in U.S. tax law of allowing a credit against U.S. income tax for foreign income taxes paid by U.S. firms doing business overseas is correct. This is a mildly controversial assumption and changing the present foreign tax treatment would affect the oil share of the energy industry.[13] The foreign tax credit, however, involves a broader tax policy issue that we cannot discuss in an energy study. We can only investigate how the general tax principle applies to the oil and gas business.

The critical features of the tax treatment of foreign operations of the oil and gas industry are these:

(1) The foreign tax credit is allowed for payment of taxes to the host countries (which are members of OPEC), even though it is dubious whether these charges are income taxes.

(2) Percentage depletion and the deduction of intangible drilling expenses are allowed for foreign operations.

The allowance of percentage depletion and intangible drilling deductions on foreign oil production is a complex problem because of the total pattern of foreign taxes on oil. Foreign taxes on oil are greater than the U.S. tax on oil would be even if U.S. tax law disallowed percentage depletion and intangible deductions on foreign operations. These payments to host countries jumped astronomically during 1973 from about $1.50 per barrel to $7 per barrel—the government take in Saudi Arabia on Arabian light 30° oil.[14]

Under our present system of foreign tax credit for income tax (or taxes "in lieu of" income taxes), credit is extended to all charges imposed by OPEC governments. Whether this should be done is doubtful. The charges in question are far heavier than the taxes paid by other businesses in the host countries.

The form of the OPEC charge is a hypothetical income and cost calculation that amounts to a flat charge without regard to actual profit. Analagous to this taxing system would be a cigarette tax of 7 cents a pack, levied whether the cigarette company makes money or not. A cigarette tax is an excise tax—just as foreign taxes levied against oil firms might be considered excise taxes—but foreign excise taxes are not allowable as credits against U.S. income tax, and so the foreign oil production taxes we are discussing are *called* income taxes.

Finally, the OPEC charge stipulates that about 20 percent of the payment to OPEC will be called a royalty, which is only a deductible cost to the oil company and not eligible for the foreign tax credit. This is a remarkably low royalty rate for the most productive oil wells in the world.

For these two reasons the position of the Treasury Department on OPEC-type charges on the foreign operations of oil companies, from the standpoint of tax theory, is doubtful. It is therefore highly appropriate to examine Treasury's position as it relates to energy policy.

As we have said, the charges imposed on foreign oil and gas in the OPEC countries are very heavy and would be higher than the

U.S. tax even if U.S. companies were not allowed percentage depletion and intangible drilling deductions. The combination of those benefits and the OPEC charge, when it is fully eligible for the foreign tax credit, is much higher than the U.S. tax. Thus, foreign tax credits are far greater than the amount allowed against income from producing oil and gas. Under certain circumstances, the excess foreign tax credit can be used to reduce U.S. tax on other foreign income. The result is that a combination of the Treasury position on the creditability of the OPEC charges, the level of those charges, and the extension of the natural-resource tax advatages to foreign operations provide U.S. oil and gas companies operating in the OPEC area with some tax advantages other than those derived from producing oil and gas.

The Excise Tax on Motor Fuel

The taxation of gasoline and diesel fuel is a special matter because federal, state, and local governments collectively provide high-way users with free highway services. We think that after a full valuation of highway costs and highway user taxes the highway user gets more service than he pays for. Highway use, which involves a heavy energy drain, is, in effect, subsidized. Thus, when a motorist buys gasoline at a filling station and pays a "tax" of 12 cents per gallon, it is more accurate to describe the situation as a purchase of gasoline with tax—plus the purchase at less than cost of a permit to use the number of miles of highway that his car will get out of a gallon of gasoline.

By and large, governments in the U.S. have decided that charging for the use of highways at the gasoline pump is more conven-ient than charging through a system of highway tolls or through parking taxes. As a result, the net subsidy is more generous for some classes of users than for others, and so policy issues arise about how and at what level highway use taxes should be imposed. These issues are com-plicated even more by the prospect that some expenditures for urban mass transit will be made out of highway taxes.

Another aspect of the gasoline tax situation is that the United States is unique among the industrial countries in not using gasoline as a net revenue source the way we use liquor and tobacco. Other countries find it useful to tax gasoline well in excess of amounts for highway expenditures.

State and Local Nonincome Taxes[15]

Apart from income and gasoline taxes, the sum of state and local excise and property taxes on the energy industry amounts to a differential or discriminatory tax on energy of about 4 percent of final sales price. This total includes local taxes on utility bills. State taxes on utility bills are, on the average, equivalent to state general sales taxes, but local taxes amount to a net differential on electric and gas energy purchases. Discrimination against energy also arises from property of privately owned public utilities being taxed somewhat more heavily than property generally, due to different assessment procedures.

State governments also tax energy industries more heavily through severance taxes, which are applied when energy resources are taken from land within these states. The effect of severance taxes on energy markets is complex because the taxes are similar to royalties and can, to some extent, result in lower royalty payments to landowners.

Direct Subsidies

The federal government is spending three-quarters of a billion dollars in fiscal year 1974 on subsidies for energy research. Better than two-thirds of this is allotted to research on nuclear generation of electricity, and about 15 percent will be spent on coal research, particularly coal gasification and liquefaction. In each of the areas research is done not just on ways to get energy but also on ways of reducing pollution related to energy.

Another energy-related federal subsidy is the program of low-interest loans to Rural Electric Cooperatives. The outstanding loans are about five billion dollars, and interest is being paid at about 2 percent. If we assume that 6 percent interest might have been charged by a non-government lender, the current subsidy comes to about two billion dollars.[16]

THE REVENUE COST OF TAX BENEFITS
FOR ENERGY RESOURCES

Petroleum

The production of crude petroleum in 1974 will be about 4.1 billion barrels,[17] which corresponds to the combined production of crude petroleum and natural gas liquids for 1972.[18] One would expect that production should rise with higher prices,[19] but we will stick with the 4.1 billion barrel figure, although it seems conservative.

The current average price for new and old oil, including imported crude, is not certain. We put it at $6.50 per barrel. The gross revenue, then, would be 27 billion dollars.

In 1970 and 1971, when the oil price was about $3.25 per barrel, the net value of percentage depletion as a tax expenditure was 15 percent. This was lower than the statutory 22 percent for a number of reasons:

1. Some wells are under the 50 percent limitation.
2. Use of percentage depletion involved loss of cost depletion.
3. Use of percentage depletion involved minimum tax in some cases.

As the price of oil rises to a figure as high as $6.50, these offsets will become less important: Cost depletion will not be appreciably higher; net income should grow faster than gross; regular tax should rise enough to reduce the importance of the minimum tax.

Somewhat conservatively we put the net value of percentage depletion at 17 percent of gross. Thus,

$$
\begin{array}{rl}
 & \$\ 4.1 \text{ bil. brls.} \\
\times & \underline{6.50} \text{ per brl.} \\
 & \$\ 27 \text{ bil. gross income} \\
\times & \underline{\ \ 17} \% \text{ value of depletion} \\
 & \$\ 4.6 \text{ bil. percentage depletion} \\
 & \qquad \text{deduction in excess of cost} \\
\times & \underline{\ .48} \\
 & \$\ 2.2 \text{ bil. tax saving}
\end{array}
$$

The most uncertain part of projection is the price per barrel. As a practical matter a change of $1 in the average price will change gross income by 4.1 billion dollars, the deduction by $900,000,000, and the tax saving by $430,000,000. This leads to the following range of estimates on the net revenue loss from percentage depletion for various oil prices (averages, domestic crude):

	(billions)
$6.25	$2.1
6.50	2.2
7.00	2.4
7.50	2.6

The December press release from the Treasury Department accompanying the initial presentation of the President's windfall-profits tax proposal indicated that the long-run supply price would be $7. With Persian Gulf oil now selling at upwards of $8, a $7 U.S. price seems too low.

By 1975 and 1976 the domestic output of oil should increase in response to the higher price. After 1976, with the opening of the Alaska pipeline, the production should jump by about one-third. At that point the revenue loss from percentage depletion should approach four billion dollars.

Natural Gas

The gross value of natural gas at the well-mouth was about 4.9 billion dollars in 1973,[20] and it should rise gradually with higher prices. We put the net value of percentage depletion at 15 percent, which yields a deduction of about $750,000,000 for 1974 and a tax saving of $350,000,000.

Coal and Uranium

The revenue loss from percentage depletion for coal and uranium is much lower. In the 1960 IRS data, the latest available, the effective rate of allowable percentage depletion for uranium was 13 percent and for coal 5 percent. (These figures were lower than the statutory rates, which were then 23 percent for uranium [now 22 percent] and 10 percent for coal [because of the 50 percent of net income limitation].) In these other fuel minerals about 10 percent of the depletion deduction was taken on the cost basis, as was the case with oil and gas. We can assume, therefore, that the relationship between allowed percentage depletion and cost depletion forgone by users of percentage depletion was the same as it was for oil—15 percent. Thus, we put the net benefit of percentage depletion at 11 percent for uranium and 4 percent for coal, on the basis of 1960 experience. The steady increase in coal prices suggests moving the coal percentage up to 5 percent on the ground that at current prices the net income limitation should not be so severe.

The value of mine output for coal in 1973 was 4.8 billion dollars, for uranium it was $173,000,000.[21] Projecting a continued rise in price and output for 1974 suggests a $130,000,000 revenue loss from percentage depletion on coal and $10,000,000 on uranium.

Intangible Drilling Costs for Oil and Gas

The most useful way to describe the revenue loss from the current deduction for intangible drilling costs is to start with the assumption that percentage depletion has already been repealed. If this were the case, the value of the deduction would amount to the difference between the current deductibility situation and capitalizing an investment expense and deducting it as the asset wears out.[22]

A problem remains about how to treat dry-hole costs. As we argued earlier, it would be logical to consider the total drilling costs of a taxpayer in a year to be the basis (or cost) of whatever oil or gas is found.

Drilling activity apparently picked up sharply after 1971[23]—a 20-percent-a year increase from 1972 to 1974. This suggests that the revenue loss from the intangibles deduction for productive wells was $400 thousand in 1974, and the loss for all wells was $800 thousand.[24]

We assume that investment for coal output will begin to grow above a replacement level in the early 1970's and that investment will grow enough to provide a 5-percent-a-year increase in output. We assume also that the "mine development costs" required for increased coal capacity will come to $5 per ton. This suggests that deductibility of development costs will be costing the Treasury about $50,000,000 in 1974. For uranium we assume that mine development costs average $1 per pound of U_3O_8 and that enough investment takes place to provide growth in capacity of 15,000,000 pounds a year. This produces a revenue loss of $5,000,000.

The level of development outlays for coal and uranium are not growing fast enough to generate an appreciable revenue loss.

Summarizing the revenue loss for 1974 we have:

Percentage depletion

	(billions)
oil	$2.2
gas	.35
coal	.13
Uranium	.01

Intangibles

	(billions)
productive oil and gas wells	$.4
all oil and gas wells	.8
coal	.13
uranium	.005
Total	$3.2 to 3.6 billion

The total is essentially a long-run estimate at 1974 levels. It describes the effect of percentage depletion and intangibles deductions by contrasting the 1974 revenue under these provisions with the revenue in 1974 if the provisions had never been in effect (other things being equal).

If these provisions were changed in 1974 there would be an additional revenue gain to the Treasury because taxpayers would have already deducted all prior intangible drilling costs and would not have any cost depletion to deduct. Thus, the Treasury revenue gain would be higher until the change was in effect long enough to build up normal cost depletion. The 1974 level of this transitional revenue gain would be $550,000,000 for intangibles and $250,000,000 for percentage depletion. It would decline about 10 percent a year.

2.2 HOW TO LOOK AT TAX DIFFERENTIALS

One common way of viewing the provisions of federal tax law that affect energy industries is to compare tax bills. Though it seems unfair that a big oil company pays a lower tax rate than an average factory worker, we would have to learn how to look at the difference before we made a judgment about fairness.[25] By the same token, the previous section presented very clinical figures on average differences among companies, but as everyone knows, an average is only an average. Some energy companies do much better than the average.

Our approach will be to suspend judgment on whether the difference in tax rates is fair. Logically, it is quite reasonable to say that a tax differential is like requiring a company to pay regular taxes, then paying the company to do such things as finding more gas and oil or reducing the price of gasoline to consumers. Congress has recognized that percentage depletion and deduction of intangible drilling and development expenses are unequal, but various committee reports are full of assertions that there are good results, such as greater national security. For this reason the provisions are referred to as tax incentives.

While we tentatively accept this approach, it raises some questions:

1. What is being sought through these incentives?
2. What results are being obtained?
3. Because a tax incentive is like a government expenditure, do these incentives have a good benefit-cost ratio?
4. Can desirable results be obtained in a better, cheaper way?

These are important questions. Because taxes on oil companies are patently unequal, there is a heavy burden on the industry to prove that the government gets value for these benefits.

In the oil and gas production business, where the largest portion of the special tax benefits are involved and where one must often encounters charges of unfairness, it appears that the rate of profit on capital after tax is not appreciably different from the rate in other businesses.[26] It is difficult to reach a final judgment on the matter because, to some extent, windfall profits can become capitalized in the capital values themselves so that the ratio of book profits to book capital will look quite normal.[27] It is also quite likely that the rate of return is better in the production end of the business (where the special tax provisions apply). This could be somewhat offset by a lower rate of return on refining and distribution.[28]

There are other complications in assessing the historical data on rates of return in oil and gas production. For example, oil and gas are commonly produced together, and since the mid-1950's the price of natural gas has been subjected to regulation by the Federal Power Commission. In the past few years, however, there has been considerable relaxation of regulation and much discussion of complete deregulation.[29] These circumstances alone make it hard to judge what the rates of return would be in a substantially different regulatory atmosphere. Another important circumstance is that, over the long run, the petroleum industry was in a prolonged slump from about 1956—following the sharp expansion after the first Suez crisis—until about 1970. This period was marked by considerable over-capacity, a condition that tends to reduce profit rates.[30]

If the petroleum industry is generally competitive, even if the competition is mostly among the majors (that is, the five largest companies), the general tendency would be for the rate of return after taxes to become very close to that of manufacturing generally.

Initially, the tax benefits would tend to generate better-than-average after-tax profits. Even with competition primarily among the majors, this would cause the existing companies to expand. In fact, it would also bring new companies into the industry. Prices would fall (or rise less than they would have otherwise) and royalties on oil and gas lands would rise. In the following section we turn to the matter of what did happen—what the actual combination of higher royalties, lower prices, increased output, and increased reserves was. In Chapter Five we deal with the issue of whether the United States should continue these policies. The question of fairness will be dealt with by asking whether there are better ways to reach our national objectives.

Before leaving the question of fairness, we should deal with some well-known statistics about the income taxes certain oil companies pay, namely, the treatment of foreign taxes. There is a lot of spectacular data about how some oil companies pay no U.S. income tax. As we have already mentioned, the general rule in our tax system is that when an American company does business and earns profits in a foreign country, income taxes paid to the host country can be subtracted from the potential U.S. tax on the overseas profits. Because foreign countries tax the income of oil and gas companies heavily, the credit for foreign taxes often wipes out the U.S. tax on foreign income. If the U.S. company had no net U.S. income, its U.S. tax will be zero. It paid income taxes, but not to the United States.

As we have already acknowledged, one can debate whether it is a wise policy for the United States to forego tax on overseas profits when the host country's tax is as high as the U.S. tax. There are also strong reasons for supporting the foreign tax credit, and, for purposes of this study, we will assume this national policy as given. At the same time, that there were some policy issues involved in specific application of the foreign tax credit to the charges imposed on oil companies by the OPEC. If we concede the principle of foreign tax credit, the magnitude of the tax paid by oil and gas companies looks very different.

An analysis of the effective tax rate problem appears in Table 2-1.

The last line is the "effective tax rate" cited by those who want to show that oil companies pay almost no U.S. tax.[31] The next to last line is the sum of the income tax paid by oil companies and the foreign taxes that qualified for the foreign tax credit. In effect, the latter figures show the U.S. tax that would have been due on foreign profits if foreign taxes had not been subtracted. The table does not show credit for foreign taxes in excess of the U.S. rate. (To give credit for the excess would be to imply that the U.S. ought to sacrifice some tax on domestic business to compensate companies for the excess taxes collected by foreign governments.) It does, however, treat excess foreign taxes as a cost of doing business abroad; so they are subtracted in figuring economic income.

The more useful indication of the income tax burden on oil companies is given in the next to last line, which indicates that the real effective rate of income tax is about 32 percent for oil and gas companies compared with 43 percent for all U.S. corporations. We think that this difference is the correct way to describe the relative tax burden of the industry as a whole. This calculation is based on the argument that only to the extent that the foreign "income" tax is as high as the U.S. tax on the foreign income should it be regarded as a tax. This is the way our foreign tax credit provisions look at it. To the extent that the foreign tax is higher than this we regard it as a cost of doing foreign business.

If one wants to regard the total foreign income tax as a tax, the figures for 1968 would be modified. The income becomes $8,061,000, the tax, after investment credit $3,082,000, and the effective tax rate 38 percent. None of these statistics, however, carry a precise meaning, because the "oil industry" is a somewhat amorphous collection of companies that engage in oil production (which qualifies for special beneifts) as well as refining, distribution, and even non-oil and non-gas businesses (which do not get special benefits). The effective tax rate on any oil and gas company depends on the relation of its oil and gas production to its other fully taxable activities.

A rough calculation applied to the 1960 data (the latest for which the income tax accounts on oil and gas production are available) suggests that the effective rate on the tax-favored part of the business may be about 24 percent.[32]

While we are speaking of effective tax rates we need to knock down the industry claim that the oil and gas industry pays a higher effective tax rate than the rest of United States business because in oil and gas the ratio of income tax to total sales is higher than the same

Table 2-1. Alternative Effective Tax Rates on Oil and Gas Corporations compared to All U.S. Corporation, 1968 (corporations with net income only)

Account	All Corporations Actual 1968 ($bil.)	Corporations in crude petroleum & natural gas	
		Actual 1968 ($mil.)	Adjusted to present law ($mil.)
Income subject to tax	81.4	4,651	5,222
plus excess depletion	4.9	2,990	2,421
plus excess of depreciation of intangibles over tax deduction	.6	420	420
less foreign tax in excess of foreign tax credit	.6	(500)	(317)
Equals economic income	86.3	7,561	7,747
Income Tax before credits	39.7	2,400	2,673
less investment credit	2.4	196	196
less foreign tax credit	3.7	1,609	1,792
Equals tax after credits	33.6	576	881
Tax after investment credit but before foreign tax credit, as percent of economic income	43%	29%	32%
Tax after all credits as a percent of economic income	39%	8%	11%

Sources: *Statistics of Income, Corporations, 1968.* U.S. Treasury Dept., Internal Revenue Service. *The Tax Burden on the Oil and Gas Industry,* Houston Petroleum Industry Research, Inc., 1972. *The Petroleum Industry's Tax Burden, Taxation With Representation,* Arlington, Va., 1973.

ratio for all business. If Congress considered this criterion relevant, it would replace the corporate income tax with a sales tax, and every industry's tax would be proportional to its sales. Congress wants a tax on the return to capital and the oil industry has a relatively high amount of capital.

Another industry claim makes the point that the oil industry pays considerable non-income taxes. The biggest non-income tax is the gasoline tax that pays for highways, which create most of the market for oil. In any case, if paying excise taxes, which are added to the final price, were a justification for income tax relief, then tobacco and whiskey companies have an even greater claim to income tax relief.[33]

2.3 THE EFFECTS OF TAX BENEFITS FOR NATURAL RESOURCES

What has the effect of income tax benefits for energy natural resources been? This is an important question about which there is obviously much disagreement.

For reasons developed below, the best opportunity to observe the effects of tax benefits is to look at the oil industry. The tax benefits for oil production are equivalent to about 20 percent of the well-head price of crude oil. About six points of the benefit goes to increase royalties and expenses. The balance serves mostly to decrease the price of crude oil, which increases the level of production about 7 percent. Possibly two points of the balance might go to oil companies.

We concentrate on the oil industry because there is more data to draw on. The natural gas industry has for nearly twenty years been subject to price regulation that takes the net tax, and thus the value of tax benefits, into account. (Presumably the benefits have served to reduce natural gas prices, but this does not tell us what would happen if natural gas prices were deregulated.) The importance of the depletion allowance in relation to the price of coal is too low to isolate the tax effect from the general depression in coal markets after the early post-World War II years. The uranium market has been influenced too much by other government policies to isolate a tax effect, and there is inadequate experience with shale.

To analyze the effect of percentage depletion and deduction of intangibles on the oil market, we start with the obvious initial effect: that tax benefits should increase the desire of companies to invest in more oil production. John McLean of Continental Oil, speaking of increased percentage depletion, said that it "would increase cash flows and rates of return, thus making new investments in exploration and development more attractive."[34]

A necessary ingredient for more exploration and development is acquisition of drilling rights on lands with oil or the prospects of oil and gas. Companies have a reserve of such rights more or less keyed to their drilling program. If the tax benefits increase the planned level of exploration and drilling, there will be an increase in the demand for drilling rights. Because the supply of land with oil prospects is limited,[35] the increase in bidding should drive up royalties.

It seems clear that one effect of tax benefits is to increase the rate of royalty payments.[36] The question is "how much?" The answer seems to be that it depends on the characteristics of the supply of drilling prospects. If there is an enormous supply of parcels with chances of a good discovery, then drillers could buy more drilling rights without driving the price up much. If the next-best prospects are much poorer, attempts to get more drilling rights will drive the price up sharply—even without tax incentives.

The critical characteristic of the supply of drilling prospects is what economists call the "elasticity of supply."[37] Davidson, Falk, and Lee have developed an ingenious analysis of the relationship between royalty payments[38] and supply elasticity. It permits an estimate of how much of the tax benefit has gone into higher royalties.[39]   Their conclusion is that the supply situation for onshore U.S. drilling is such that the landowner share of net rents (that is, income in excess of drilling cost, including a normal return on capital) is about 25 percent; for offshore drilling on the U.S. outer continental shelf the landowner share is close to 40 to 50 percent.

Because, initially, the tax benefits amount to a net rent to producers, we can estimate that roughly 35 percent of the tax benefit goes into higher landowner benefits. This much of our result is still consistent with a net addition to the rate of drilling and exploration. In fact, royalties only rise because the tax benefits increase drilling programs.

Rising royalties, however, absorb some of the incentive effect of the tax benefits and serve to reduce the growth in drilling.

The way in which Davidson, Falk, and Lee develop their analysis permits us to test their calculations from royalty records in an independent way. Their analysis shows that the landowner share of the net rent of 40 to 50 percent implies an elasticity of supply of 1.3 to 1.0.[40]

We can cross-check their estimate by looking at other estimates of the elasticity of supply; other current estimates put the supply elasticity, measured from past price and output behavior, very near 1.0. Erickson estimated the elasticity at 0.87.[41] Steele put it at 0.8,[42] and Steele interprets estimates developed by the Humble Oil Company as implying an elasticity of 1.0.[43]

At a very basic level one could be satisfied with the conclusion that the elasticity is "about 1.0" and that the share of net rents, and hence of tax benefits, going to royalty recipients is "about 50 percent." The Davidson-Falk-Lee data, however, do show a persistently lower royalty, especially for onshore drilling. This can most easily be explained by assuming that the bargaining situation between landowners and the oil companies is such that the oil companies can retain some of the net rents. This interpretation is strengthened by the low royalty share for U.S. onshore drilling—only about 25 percent, which would imply an elasticity as high as 3.0. Bain, looking at Pacific Coast properties, concluded that if landowners had been able to extract all of the net economic rents, their share would be 48 percent, which implies an elasticity of 1.1. The obvious explanation for this discrepancy is that in an actual market situation many landowners are dealing with a small number of drillers, and the drillers are in possession of better information on oil prospects and oil markets. The landowners, then, get a lower share than they would have gotten in a perfect market.[44]

This explanation of the discrepancy between royalty data and independent supply elasticity estimates, which attributes the discrepancy to the bargaining position of the oil companies, is consistent with the different royalty shares for onshore and offshore drilling. If Bain is right, the onshore royalty shares should have been as high as the offshore ones.[45] When offshore properties are involved, the lessor is the federal government, which can set up reasonably sophisticated bidding procedures.

Our theory that oil companies retain some of the economic rent is supported by other statistics. McDonald's data shows, for example, that in the production business oil companies earn better than normal profits.[46] This evidence leads us to conclude that the benefits of percentage depletion are divided in the following way: 40 percent to increased royalties, possibly 10 percent to increased after-tax profits of oil companies, 50 percent to price reduction.

The evidence so far has been concerned only with the effect of tax benefits on royalties. If half of the tax benefit were to go into more drilling, the oil supply would be increased, which in turn would reduce the price of oil. The expansion-of-drilling effect would stop when the supply increased enough to push the price down by 10 percent, which would just offset the drilling incentive part of the tax benefit (which we concluded was half of the 20 percent tax benefit).

The question now is, "What is the demand relationship between price and quantity, and how much can sales of oil expand before the price falls by 10 percent?" The best available answer is 3 percent, which is implied by the commonly accepted estimate that elasticity of demand for crude oil is 0.5.[47] The 10 percent price reduction for crude oil, then, should increase production by 3 percent and, in the long run, increase drilling by 3 percent to provide more reserves to replace those used up by more production.

This analysis of the effect of tax benefits follows a rather specialized theory of behavior on the part of oil companies with respect to reserves. Specifically, it assumes that the effect of the tax benefits is, in the first instance, to reduce the cost of capital investment by oil companies; the reduction should lead to additional investment.[48] If we think of investment in oil wells as following ordinary business investment practice, we would expect oil production operations to become more capital intensive, that is, oil companies would carry more reserves of oil in the ground. The assumption of the statistical analysis presented in this section is, however, that oil companies do not behave in this way. By and large, more investment in response to tax benefits does not lead to more reserves than output; instead, it leads to more output and only a proportionate increase in reserves. In Chapter Five we will develop in detail the strong reasons for believing that this is the correct way to look at oil investment decisions.

We need to add a qualification for an uncertain future; however. The analysis was built on past experience, which included an oil import quota. In the near future it is plausible that the world import price will set sort of a ceiling on the domestic market oil price. (Domestic well-head prices and landed prices of imported oil of the same grade would still differ from each other by transportation costs.) We can only speculate about the situation. Probably, 40 percent or so of tax benefits would still go to royalties, but there would be less price reduction and the production increase would be closer to 10 percent.

NOTES TO CHAPTER TWO

[1]A firm producing and refining oil may make an income of 200 from its total operation, 100 on producting crude oil and 100 on refining it into gasoline, fuel oil, and so forth. Percentage depletion applies only to the income from producing crude oil. When the firm refines its own crude, the income from crude is a hypothetical calculation involving the costs of producing crude and the market value of the crude produced.

[2]If a coal firm mines and sells coal for 100 and has mining costs of 88, then it cannot deduct the full 10 percent of gross income as percentage depletion; that is 50 percent of 100 less 88 is 6 percent of gross income.

[3]A detailed derivation of these estimates is provided in G. Brannon, "The Present Tax and Subsidy Provisions Relating to the Energy Industries," *Studies in Energy Tax Policy,* G. Brannon, Editor, Cambridge, Mass., Ballinger, 1974. This volume is hereafter referred to as *Studies.*

[4]Intangible drilling expenses, which are about 47 percent of drilling costs, are basically the labor and material costs incurred in drilling. They do not include the cost of drilling machinery (tangible) that can be used again. The major deductible expenses for hard minerals are the costs of developing the mine.

[5]Using a 20–year life of the asset, a 10 percent discount rate and the double declining balance method, the current value of the depreciation deductions over the life of the asset would be half the value of current deduction.

[6]There is a fuller discussion of these effects of the tax provisions along with some estimates in section 2.3 of this chapter.

[7]See G. Brannon, *op. cit. in Studies.*

[8]Susan Agria, "Special Tax Treatment of Mineral Industries," *The Taxation of Income from Capital,* A. Harberger and M. Bailey, Editors, Washington, D.C., Brookings, 1969, pp. 77–122.

[9]The comparison offered here is directly applicable to relative income tax benefits for coal, oil, and gas in the heating market. No precise comparison can be made for markets that are unique to one fuel, such as coke or gasoline. We think that the general implication of the electricity calculation is relevant to the total markets for oil, gas, coal, uranium, and shale. It just happens that we can compare the various fuels better in the electricity example.

[10]G. Brannon, *op. cit. in Studies.*

[11]These relative capital costs were obtained from data in *Statistics of Income, 1960 Supplementary Report on Depletion Allowances,* Internal Revenue Service, p. 37. For both coal and oil-gas, we assumed a 50 percent tax rate and used the depletion deduction plus 50 percent (net income before depletion less depletion) for the numerator and gross income from the mineral properties for the denominator.

[12]More detail on these and other tax provisions discussed in this chapter is contained in the addendum to Chapter Five.

[13]For a general discussion of both sides of the controversy, see Peggy Musgrave, *Issues in the Taxation of Foreign Income,* Cambridge, Mass., Harvard University Press, 1970. Most tax economists come out with more sympathy for the tax credit than Professor Musgrave expresses.

[14]*Business Week,* January 1, 1974, pp. 19, 20.

[15]This section draws on Allen Manvel, "A Survey of the Extent of Unneutrality Toward Energy Under State Excise, Property and Severance Taxes," in *Studies.*

[16]Based on data in the appendix to *The Budget of the United States Government, Fiscal Year 1974,* p. 165.

[17]Fact sheet from White House accompanying the energy message, January 23, 1974.

[18]*Survey of Current Business,* November 1973, pp. 5–35.

[19]*The Oil and Gas Journal,* December 24, 1973, reports considerable activity to increase production.

[20]*Oil and Gas Journal,* January 24, 1974, p. 35, citing U.S. Department of Interior.

[21]*Oil and Gas Journal, ibid.*

[22]If the deduction of intangibles were disallowed and percentage depletion continued, the revenue gain to the Treasury would be greater because the deduction as the asset wears out would be cost depletion, which would be lost if the taxpayer took percentage depletion.

[23]The *Oil and Gas Journal,* December 17, 1973, p. 39, reports an increase in drilling costs of 20 percent between 1971 and 1972.

[24]The basic calculation assumes that there would be no revenue loss if intangibles stayed at 1.3 billion dollars for productive wells and three billion dollars for all wells after 1971. The net revenue loss was calculated by assuming the 20 percent annual growth for each category and finding the difference between current deduction of the growth element and the cost depletion deduction in 1974 under 20–year double declining balance write-off for the growth elements from 1972 to 1974. The growth element each year is cost over 1.3 billion dollars and three billion dollars respectively.

[25]For considerable data of this sort cf. Phillip Stern, *The Rape of the Taxpayer,* New York, Random House, 1973, and testimony of Congressman Charles Vanik before the Joint Economic Committee, July 1972. *Panel on Tax Incentives.* Congress expects large corporations to pay tax at about a 48 percent rate and ordinary wage earners to pay at an effective rate of about 10 percent.

[26]*Petroleum Facts and Figures,* Washington, D.C., American Petroleum Institute 1971, p. 513. Data attributed to the National City Bank.

[27]To illustrate this, we assume that a business is earning a rate of profit of 30, or 30 percent, on a capital investment of 100. If this firm is sold, it will probably be for about 300, in which case the buyer would be getting a normal return of 10 percent (30 on 300). After the sale the windfall profit is still there, but the statistics would show the new firm as earning only 10 percent on its capital, that is, its purchase price.

[28]Stephen McDonald, *Federal Tax Treatment of Income From Oil and Gas,* Washington, D.C., Brookings, 1963, Appendix A.

[29]See *Regulation of the Natural Gas Producing Industry,* Keith Brown, Editor, Resources for the Future, Baltimore, Johns Hopkins University Press, 1972.

[30]Stephen McDonald, *Petroleum Conservation in the U.S.: An Economic Analysis,* Resources for the Future, Baltimore, Johns Hopkins University Press, 1971, p.2.

[31]Testimony of Hon. Charles Vanik in *Tax Subsidies and Tax Reform,* Hearings before the Joint Economic Committee, July 19, 1972.
Sources: *Statistics of Income, Corporations, 1968.* U.S. Treasury Dept., Internal Revenue Service. *The Tax Burden on the Oil and Gas Industry,* Houston Petroleum Industry Research, Inc., 1972. *The Petroleum Industry's Tax Burden,* Arlington, Va., Taxation With Representation, 1973.

[32]Net income from mineral production, before depletion was $5,107,000. We reduce this to obtain economic income: minus $1,465,000 for deductions on nonproducing properties; $279,000,000 for cost depletion; $361,000,000 for 16 percent of percentage depletion, the part that we think is equivalent to cost recovery. We also raise the result by $200,000,000 for the difference between the deduction of intangibles and the depreciation of intangibles, a number calculated from the relationship of this difference to percentage depletion in 1970 based on *The Tax Burden on the Domestic Oil and Gas Industry,* Houston, Petroleum Industry Research, Inc., 1972. This produces economic income of $3,202,000. The tax base was derived as the same net income reduced by allowable depletion and deduction on non-producing properties, which was $1,112,000. A rate of 48 percent on one-third of economic income produces 16 percent. Adjustment for the 1969 change in the percentage depletion rate and the minimum tax would raise this to about 24 percent. Data from *Statistics of Income 1960 Supplemental Report Depletion Allowances,* U.S. Treasury, Internal Revenue Service.

[33]The basic industry discussion of relative tax burden is in *The Tax Burden on the Domestic Oil and Gas Industry, op. cit.* Further discussion of these claims is in *The Petroleum Industry's Tax Burden,* Arlington, Va., Taxation With Representation, 1973.

[34]Testimony in *General Tax Reform,* panel discussions before the Committee on Ways and Means, U.S. House of Representatives, Part 9, February 26, 1973, p. 1,246.

[35]Over time, new discoveries may reveal hitherto unknown prospects, but we are investigating the initial response to the tax incentive, and in this period the prospects are fixed.

[36]Paul Davidson, "Public Policy Problems of the Domestic Crude Oil Industry," *American Economic Review,* 54, June 1964, p. 303; D.R.G. Campbell, "Public Comment," *American Economic Review,* 53, March 1964, p. 114; Thomas Stauffer, "Estimated Economic Cost of U.S. Crude Oil Production," paper presented to Society of Petroleum Engineers, Fall Meeting, October 8–11, 1972.

[37]Elasticity of supply refers to the ratio of the percentage increase in output to the percentage increase in price. Thus, an elasticity of 1.0 means that a 1 percent increase in price brings about a 1 percent increase in output, other things being equal.

[38]The reference to royalty payments here and in the following discussion includes total payments to landowners including royalty rates, lease bonuses, shut-in payments and so forth. Payments before production are converted into equivalent royalty rates by taking into account the time value of money, that is, the discount rate. A payment of $1,000 before production is in this sense more than an additional payment of $1,000 in royalties proper spread over 20 years of production. These considerations explain why the following shares of "royalty payments" seem to be much higher than the typical royalty rate of one-eighth.

[39]See Davidson, Falk, and Lee, "The Relations of Economic Rents and Price Incentives to Oil and Gas Supplies" in *Studies.*

[40]A further explanation of this result is offered in Chapter Four.

[41]Edward Erickson, "Economic Incentives, Industrial Structure and the Supply of Crude Oil Discoveries in the United States, 1946–1958/1959," unpublished Ph.D. thesis, Vanderbilt University.

[42]This is implicit in the supply price relationship presented by Steele in a statement in *Trends in Oil and Gas Exploration,* Hearings before the Committee on Interior and Insular Affairs, U.S. Senate, August 8 and 9, 1972, Part I, p. 254.

[43]*Ibid.*

[44]Davidson, Falk, and Lee, op. cit. in *Studies,* and J.S. Bain, *The Economics of the Pacific Coast Petroleum Industry,* Vol. II, Berkeley, University of California Press, 1945.

[45]The analysis we are citing is an historical one. Today the royalty rates on good offshore properties should be higher than for onshore properties. The Bain argument is that in their day the Pacific Coast onshore properties should have commanded as high a royalty share as offshore properties can command today. Also, the analysis does not apply to windfalls to lessors who paid for drilling rights when the oil price was around

$3.50 per barrel and are able to enjoy prices of $7 or more.

[46]S. McDonald, *Federal Tax Treatment of Income From Oil and Gas, op. cit., Appendix A.*

[47]J.C. Burroughs and T.B. Domenich, *An Analysis of the United States Oil Import Quota*, Lexington, Mass., Heath, 1970.

[48]For some dimension on tax benefits as reductions in capital cost and as incentives to more capital investment, see Susan Agria, "Special Tax Treatment of Mineral Industries," in *The Taxation of Income from Capital*, Harberger and Bailey, Editors, Washington, D.C., Brookings, 1969; Peter Steiner, "Percentage Depletion and Resource Allocation," in *Tax Revision Compendium*, Committee on Ways and Means, U.S. House of Representatives, Vol. II, 1959, pp. 949-966; and Arnold Harberger, "The Taxation of Mineral Industries," in *Federal Tax Policy for Economic Growth and Stability*, Joint Economic Committee, 1955, pp. 439-449.

Chapter Three

An Overview of the Adequacy or Defects of Energy Resource Markets

3.1 DO WE WANT MORE RESOURCES FOR ENERGY?

This question is at the heart of why many people find it hard to understand how economists deal with policy issues. The offhand answer would be, "Of course we want more resources for energy. There's an energy crisis."

Despite the magnificence of American productivity, we cannot, in general, produce more of everything at any given time.[1] If we rearrange our productive capacity to produce more resources for energy, then we will be producing less of something else. What we give up to get more resources for energy is the cost of those resources. The economist, then, answers the basic question: "Only if more resources for energy are worth the cost."

As we said in our introductory chapter, there is a simple way of getting initial information about whether more energy resources are worth the cost. The value of turning more of our land, labor, and capital to producing energy is worth the price if there are some productive factors now devoted to producing such items as food or appliances that could make more income by producing energy.

The cost of more energy is the potential loss of output of other things. Unless productive factors can be utilized better by producing energy rather than other items, then the answer to the question that headed this chapter is, "No, we don't really want more resources for energy."

The preceding discussion demonstrates the operation of the ideal price system that we talked about in Chapter One. The problem is, of course, that there may be defects in the price system, so that the actual operation of the market will produce less energy than we really want. If the defects are serious, the actual production of energy might be so far below what we want that it is called an "energy crisis."

The defects in the price system relevant to the energy crisis are brought up in different contexts throughout this report. In Chapter Seven, for example, which discusses taxes and subsidies for electric utilities, we analyze defects arising from the ways in which public utility prices are regulated. In Chapter Eight, we discuss the defects arising from the fact that some energy producers are not charged for the environmental damage resulting from their operations. In this chapter, our concern is with defects in the markets that involve the production of natural resources that can be converted into energy.

On the face of things, energy markets are unique, and it is not unreasonable to expect that there will be "defects" or distortions from ideal operation. Most of the resources for energy are fossil fuels available in the earth's crust in unknown quantities, and finding them is expensive and risky. Our knowledge of available sources and our ability to find new sources are quite limited. We must always assume that there is considerable chance of their "running out" or becoming much scarcer than they are now. Since 1918 Congress has been convinced that if the market were left to itself there would not be enough investment in the production of petroleum and natural gas. In effect, then, Congress has consistently stuck to the view that there is a serious defect in the market price for energy resources, particularly oil and gas, that needs to be offset by tax incentives.

In this chapter, we will explore in depth several potential peculiarities of the oil and gas industry that might indicate that the price system would not work satisfactorily on its own and that its operation would be improved by the specific intervention of taxes or subsidies.[2] There are several reasons why the oil and gas industry might be considered different from other industries:

(1) The United States income tax on corporate profit may bear especially heavily on the oil and gas industry, leading to inadequate investment.

(2) The risk and uncertainty in the oil and gas industry may result in a level of investment that produces inadequate reserves to deal with future uncertainties.

(3) The special activity of searching for a hard-to-find resource, such as oil or gas, may involve external benefits and costs that lead to too much or too little investment.

Special tax incentives were applied first to oil and gas and not extended to coal until many years later. The uncertainties about oil and gas supplies have seemed greater than corresponding problems in other resource areas. For these reasons, it seems appropriate to explore the basic price-defect issue in the context of oil and gas.

Serious attention has been given to the problem of whether existing markets could meet the financial requirements of the U.S. energy industry. In the spring of 1973 a Senate Committee held public hearings on just this topic.[3] Richard Gonzales, a consulting economist previously associated with Humble Oil (now part of Exxon), told the Committee that continuation of the present differential tax treatment was "appropriate and effective in attracting needed capital into this risky business that is also capital intensive."[4]

To evaluate the argument that tax incentives are needed to raise the proper amount of capital, we must assume that capital markets are reasonably efficient and then ask whether there are defects in these markets that will create trouble in financing the oil and gas share of the energy industry.[5]

In an ideal capital market there is a certain amount of capital for which various borrowers compete. If the economy "demands" more oil and gas than is being produced, a measure of the demand would be a price that makes the oil and gas business highly profitable. Out of the higher profit oil and gas companies would be able to offer sufficiently good prospects (or pay high enough interest rates) to raise the capital to expand output. When the price of oil and gas gets so high that consumers are not willing to pay more for oil to enable oil companies to acquire a still larger portion of the available capital, then the amount of capital going into the oil and gas industry would be at the ideal level.

3.2 THE ASPECT OF THE CORPORATE INCOME TAX

One major characteristic of the capital market in the United States is that our tax system bears rather heavily on corporations. There is an element of double taxation in our system of taxing corporate profits and taxing the realization of these profits by the owners in the form of dividends. We will not debate here whether it is a good idea to tax corporations this way; the taxation is a fact. We do, however, need to examine the issue, raised by Professor McDonald, of whether this kind of corporate tax would result in particularly heavy over-taxation of the oil and gas industry if the tax were not moderated by such tax benefits as percentage depletion.[6] The argument that a uniform corporate tax would create a special burden on the oil and gas industry has been conducted primarily in economic journals, but it has some obvious parallels with the Gonzales contention that without special tax provisions there would be inadequate capital for the oil and gas industry.

One must concede that corporate tax distortion cannot be an argument for differential treatment in favor of the oil and gas industry if the burden of the tax is carried by all corporate capital. If the tax merely lowers the rate of return on capital, then *all* "borrowers"[7] in the capital market will get the lower returns and no particular borrower is discriminated against.

The situation is different if the corporate tax is "shifted," that is, compensated for by charging higher prices. In this case there is more tax to be shifted in industries in which the ratio of profit to price is high. McDonald concluded that, because of risk and capital intensity, a high rate of return is characteristic of oil and gas production. This would mean that a tax that was a uniform percentage of profit in all industries would increase the price of oil and gas more than the price of other goods and services. With this higher price, sales and output of oil and gas would be reduced relative to other outputs—in effect, a market defect. The implication is, then, that this distortion needs to be offset by some tax benefit, such as percentage depletion, to reduce the corporate tax burden on this industry.[8]

Very simply, the kind of argument that one might develop for percentage depletion runs like this. If the corporation tax is shifted into prices, then a 15 percent price increase would be required in an industry where the profit rate is 15 percent, and a 5 percent increase would be required where the profit rate is 5 percent.[9] The tax increases prices more in the high-profit industries than in the low-profit industries.

Several strong counters to this argument for percentage depletion cause us to reject it. One deals with the shifting of the corporate tax, and the other deals with the assertion about profits in the oil industry.

The McDonald argument depends, in the first instance, on the assumption that the corporate tax is shifted forward in prices because only in this way can it distort the oil *price*. We think that the corporate tax is not significantly shifted into higher prices.[10] Even if the corporate tax *is* shifted into prices and thus serves to increase the price of some goods more than others—that is, to distort prices—the appropriate response would be along either of two lines. One is to accept a shifted corporate tax for what it is, a more or less deliberate decision to penalize capital, which implies that we should be satisfied with the "distortions" that are no more than a lower penalty on products that require much capital. The other response, if we are dissatisfied with shifted corporate tax, is to change the corporate tax. In neither case does it make sense to keep the corporate tax, but to reduce it for only one kind of business.

Another reason for discounting the shifted-tax argument for percentage depletion is that the rate of profit for the oil and gas industry taken as a whole is not so different from that in other industries—although it is relatively greater in the production part of the business than in refinery and distribution. It is very plausible that the high rate of after-tax profit is not explained by the peculiarities of this kind of business but is due to the tax benefits themselves. There is considerable reason to believe that risky industries will show, on the average, not very different rates of return than relatively safe industries. While the occasional high profit in a risky venture is taxed accordingly, the loss incurred will usually create a proportionate tax saving when it is deducted from some other income (or carried back or forward to income in another year).[11]

3.3 THE ASPECT OF RISK

Another reason commonly given for generous tax treatment of income derived from oil and gas investment is that these businesses are, as mentioned before, very risky.[12] This argument is not concerned with the effect of a shifted corporation tax on the price of oil and gas, but is based on the view that because of risk the oil and gas industry needs a higher rate of return, and that unless the tax law provides some special relief there will be too little investment. To deal with the risk aspect, we must identify several kinds of risk and comment on the usual ways in which they are handled in the industry.

First, there is technological risk. An oil-well driller may drill a dry hole. A movie producer may release a flop. An automobile manufacturer may have to recall a car model. There is the possibility of loss in operating any business. Because we need to get a handle on what it means to say that a business is "very risky," we will define the term as a high probability of losses that could wipe out a business.

Following this definition oil and gas production seems risky because the probability of success in wildcat drilling is quite small. About one in nine exploratory wells is a producer; about one in fifty is commercially profitable.[13] It is plausible that a driller could lose a lot of money.

This is not the case, however, because of a simple statistical matter called the "law of large numbers." According to the law, the likelihood of the percentage of unfavorable results differing from the expected percentage declines as the number of repetitions of the event increases. Thus, on drilling ten holes the probability of no discoveries at all is about 30 percent. On drilling 100 holes, the probability of no discoveries falls below .001. With 100 holes the probability of having three or more fewer successes than the expected number of successes, eleven, is only 30 percent. In relative terms, risk is reduced with more independent repetitions of the risky event.

A large oil or gas company has an opportunity to make many drilling decisions. Some will be successful, and some not. For drilling to be profitable in the long run, we must assume that the value of successful wells must more than cover the total cost of both successful and unsuccessful drilling. The risk for any company is associated with its having less than the expected—average—success. As in any business, this is possible; but for a relatively large company, the probability of a relatively large loss becomes small.

An oil or gas company does not even have to be large for the law of large numbers to apply. All the company needs is a large number of risks. There are many devices for spreading risk in well operation, such as selling partial interests in some wells to get money for drilling more.[14] Despite the proclivity of oil witnesses before Congress to cite the low success probabilities for a single wildcat drilling operation, success probabilities on single wells are not a useful description of the risks facing a whole company.

It is plausible, however, that there are more substantial risks in the oil and gas industry than wildcat drilling. There are several possible sources:

1. Changes in markets due to the discovery of big new fields, such as the North Slope in Alaska.
2. Changes in the market created by delay in developing an anticipated source, for example, the stalemate between government and environmentalists over where to put the Alaskan pipeline.
3. Changes in gasoline demand due to regulations on new automobiles.
4. Changes in oil import quotas and tax laws.
5. Development of new ways to use coal.
6. A fantastic price increase engineered by the international cartel.

Unlike the technological kind, these risks are not affected by the statistical law of large numbers. If an oil company is planning expansion in, say, California over the next five years, then there is only one North Slope with a pipeline that might pour oil into the California market in that period.

In some businesses there are techniques for dealing with unique risks, such as the futures market. A miller who wants to contract to sell his year's output of flour may be concerned that something unexpected will change the price of wheat. As a hedge, he can buy futures. A shipowner who is taking a high risk on an unusually valuable cargo can buy insurance.

In the oil industry there are no organized futures markets or insurance possibilities (beyond the risk-sharing in wells described earlier), but this is characteristic of most manufacturing.[15] What is unique to the oil industry is that disturbances may not be self-correcting. In most businesses, even if there is an unexpected new source of supply analogous to the North Slope, one can expect the established firms to reduce their normal investment; then, in a few years, everything will be back to normal. The new supply would initially lower prices. As existing firms reduce output, cut back their investment plans, or even divert some of their existing facilities to different activities, the over-supply is moderated, the price recovers somewhat, and producers are saved from financial disaster.

The normal adjustment process may not always take place, however. The internal dynamics of the industry may lead individual firms to try to cut losses in the face of over-supply by increasing (not decreasing) their output. This can make the total industry worse off as the price is driven down further. Inverse reaction to over-supply has frequently been the situation in agriculture.

If such an inverse reaction were to take place in the oil industry, normal business risks could be converted to very large losses, and we would call the industry "very risky."

In his investigation of the reaction problem in the oil industry Stiglitz concludes that the market for oil is likely to be fairly stable: in the face of disturbances offsetting tendencies arise and risks are contained.

Supplementing the Stiglitz analysis is the structure of state conservation regulation, which in the past served to protect the industry from a tendency to over-produce in the face of over-supply. For instance, the Texas conservation statute to prohibit waste, which includes "the production of crude petroleum oil in excess of . . . . reasonable market demand," was adopted as a reversal of previous policy in 1932.[16] In addition to the problems of the Depression, the Texas legislature had a fresh memory of spectacular over-production in the newly discovered East Texas field in 1930.

If businessmen were neutral with regard to risk in their investment decisions, then there would be no difference between a risky investment that had a 50–50 chance of returning a 20 percent profit or a zero profit compared with an investment that had an assured return of 10 percent. On the other hand, if businessmen were averse to carrying risk, it would show up in not investing in 50–50 deals unless the prospects were, say, a .5 chance of a 22 percent profit and a .5 chance of a zero profit. Under this risk-aversion assumption we would expect investors to show up with 11 percent average profits, the extra point being a compensation for risk.

There is some evidence on the risk situation in oil, namely, the industry argument that the rate of return on investment in the oil and gas industry is no higher than in the average manufacturing industry. The implication is either that for typical oil companies total risk is no different from average, or that oil companies are quite indifferent to risk and look only to average expected return.

If one still insists that risk is, on balance, higher in the oil industry, how does this argument apply to present tax? In the first place, a measure such as percentage depletion is particularly inappropriate for the problem of risk. As we argued in Chapter Two, part of the effect, historically, of percentage depletion is to increase the supply of oil, which decreases the price, and to increase royalties, which are costs to the driller. Both of these results increase the losses for an unsuccessful firm and increase the rewards for the successful firm. Percentage depletion is a percentage of success, that is, of increased income; the more success, the more tax benefit.

The strategy implicit in the present tax law to benefit winners in the oil game and penalize losers increases risk. A neutral tax law to some extent reduces risk by allowing net losses from unsuccessful ventures to be deducted against other taxable income. If one really believed the argument that the special risks of the oil business required special tax are not freely reproducible. For this reason, the income earned from owning land is different from the income earned by owning a machine.

Another argument has to do with whether risk should be offset at all. Stiglitz examines the argument in detail and concludes that the net risk, if it exists, is a real cost that should not be offset. Presumably, the wages of coal miners are higher to compensate them for the unpleasantness and risk of working in a mine. Because this makes the price of coal higher, we should leave it to the market to decide whether buyers want coal enough to pay people to assume risk.[17] Risk in oil business is no different.

3.4 THE PROBLEM OF RESERVES

Another way in which the oil industry may be different from other industry and thus deserve special tax allowances is in the matter of reserves.

A business firm ordinarily expects a certain level of sales, but it carries inventories to take advantage of higher demand if it should arise. Firms with large inventories can make a lot of money if a shortage develops in the absence of effective price controls.

In the typical business, carrying extra large inventories has a cost. It ties up capital that could be used for other things. Businesses can, and sometimes do, carry too much inventory. In fact, a market economy relies on the profit motive to give business firms the appropriate signals about how much inventory to carry. In the light of how likely it is that extra inventory can be sold at good prices, the firm will decide how much of its capital to tie up in inventories rather than, say, in modernizing its equipment.

There is no reason to expect that government could improve things by issuing orders to a firm about how high its inventory should be. It is common, however, for oil industry spokesmen to assert that there is a special problem of reserves that justifies such programs as special tax incentives to encourage firms to hold larger reserves.

We can concede that the reserves problem may be different for oil. For one thing, some of the situations that call for reserves, such as national security emergencies, are likely to involve price control and rationing. Oil companies would be precluded from profiting on reserves in this eventuality, so it will not be profitable for them to prepare for the eventuality by tying up funds in what (1) would not be a profitable venture even if the emergency developed and (2) would just be an extra cost if the emergency did not develop.

There is another dimension to the reserve problem associated with oil emergencies that could be called a national security problem. Individual companies should have less accurate knowledge than the government about the political factors that underlie the probabilities of national security reserves being needed.

One must be careful with this argument. While governments should be in the best position to evaluate security problems, they may make grievous errors—underestimates (Europe in the 1930's) or overestimates (the alleged missile gap that heavily influenced U.S. policy in the late 1950's). After the warning, we can, in the context of the present study, assume that government might properly be dissatisfied with the performance of private markets in providing reserves.[18]

It is not easy, however, to identify what kind of policy is needed to deal with inadequate oil reserves in private markets. For one thing, tax benefits, such as percentage depletion and deduction of intangible drilling expenses, are apt to increase output, reduce prices, and increase consumption. These effects alone provide some short-run security by increasing domestic output, reducing imports, and thus reducing the bad effect on U.S. markets of an interruption of imports. In the long run, the policy of increased reliance on domestic markets means a more rapid rate of exhaustion of the best U.S. resources, which makes the policy somewhat self-defeating. Using up the best domestic resources faster will reduce the future ability of the domestic industry to compete with imports.

Granted that there is a security reserve problem, it is still not easy to see what to do about it. We turn to the practicalities of coping with it in Chapter Five.

3.5   THE ASPECT OF EXTERNALITIES OF DRILLING

One other aspect of the oil and gas industry is the particular way in which producing firms get oil and gas reserves. They can do a c rtain amount of educated guessing on the basis of surface geologic indications, but the surest way of finding whether there is oil or gas under a plot of land is to drill for it. A particular drilling operation may yield oil or gas, but another important yield is information.

When a wildcat driller drills on a new piece of land and strikes oil or gas, he gains important information about the existence of a geological trap that may be extensive enough to interest owners of other tracts in the area in the considerably higher probability of finding oil or gas under their land. Similarly, a series of dry holes may be important negative information for other drillers in the area.

As an indication of the market value of information generated by a successful strike, we can consider the increase in lease bonuses that might be paid by drillers on surrounding tracts. A striking illustration is the one following the ARCO discovery of oil at Prudhoe Bay in Alaska, where lease bonuses on surrounding land jumped from $9 an acre to almost $2,200 an acre.[19]

To clarify the significance of this, some simplified hypothetical figures are shown below.

Drilling Program	(1) Cost of drilling	(2) Expected value of reserves discovered	(3) Expected increased bonuses on nearby tracts
A	100	150	50
B	100	120	40
C	100	80	30

The table shows three increments of drilling, each of which costs $100. To the common-sense notion that the best prospects are drilled frist, we add that the discounted expected value of reserves are

150, 120 and 80, respectively. If we compare columns (1) and (2), it will be obvious that a driller should only undertake Programs A and B. In each case, the expected yield to the driller is greater than the cost of drilling. Expanding drilling to cover Program C would be inefficient because the expected yield is below the cost.

Taking other things into account, however, Program C is worth more (80+30) than its cost (100) and should be undertaken. The 30 points of extra value in Program C—increased lease bonuses on surrounding tracts—brings the total worth of the program above its cost. The lease bonuses are amounts that other drillers are now willing to pay in view of the valuable information they got as a by-product of the drilling. If our hypothetical figures were actual figures, they would tell us that the free market would undertake only drilling Programs A and B and that Program C would be ignored.

The figures provide a highly simplified example of the basic problem of drilling externalities. In the real world many kinds of arrangements are made to reduce externalities.[20] A common one is for a driller to buy up leases on surrounding tracts. Another one is for owners of surrounding tracts to share some of the drilling cost, for example, by agreeing to pay some of the cost if the well is dry.[21] These arrangements only reduce, not eliminate, the externalities. There is, then, a case for some policy, such as a subsidy, to get drillers to do more drilling, that is, up to level C.

In the process of drilling for oil and gas reserves another externality must be considered. In the earth's crust there is an exhaustible amount of recoverable oil and gas. When one driller finds, say, a million dollars' worth of oil in a new location, he does not add a full million dollars' worth to the world's wealth. One way of describing the event is to say that the driller moved discovery up in time. Another is that the one successful discoverer made life a little harder for all the other oil and gas prospectors by reducing their chances of finding oil and gas.

A homely example will help explain this. If a person loses a ring worth $500 in his backyard, he may offer some boys a reward for finding it. The reward will be considerably less than $500 even though getting the ring back for any price up to $499 would be cheaper than buying a new ring. The point is that the ring is there. The boy who finds the ring does not really add $500 to the world's wealth, he simply gets the ring back into circulation faster. The person who lost the ring might eventually find it himself during a backyard barbecue. In oil and gas, however, the "rule of capture" is equivalent to a rule of finders-keepers, which amounts to over-rewarding the finders.

Both Stiglitz and Peterson deal with the externalities inherent in the drilling process. Both conclude that information is an important externality, partly because it is very hard for an exploratory driller to assemble large blocks of land prior to drilling. About one-third of the known oil fields cover 40,000 acres, but typical lease blocks assembled prior to exploratory drilling average about 9,000 acres.

Another aspect of the information problem is related to the federal government's practice of leasing onshore properties in the form of small leases on a non-competitive basis. There are often hundreds of applicants for desirable parcels, and successful applicants are drawn by lot. This is a highly inefficient scheme because it creates small, divided leases and means that most of the long-term benefit from a successful well will accrue to persons other than the driller. One policy change that seems to be called for, in this case, is an effort to facilitate the assembly of large lease blocks in unexplored areas. This could be done by competitive bidding.

Another approach to the problem is to get the government into the business of generating information as pointed out by Miller.[22] One of the costs associated with oil drilling is making geological and geophysical studies aimed at evaluating oil and gas prospects prior to drilling. It would be sensible for the federal government to compensate generously for these oil and gas exploration costs rather than to provide favorable treatment for the drilling itself or for production from known wells. The government might even undertake these studies itself and make the information available to the public. Peterson recommends such a positive information policy to the extent of suggesting that the government might do some exploratory well drilling itself.[23]

A third line of policy the government might follow with regard to exploration would be to focus quite explicitly on non-tax subsidies associated with exploratory drilling. This approach involves distinguishing between exploratory and development drilling, and the nature of the problem suggests that it would be sensible to draw a line that clearly excludes development drilling. For example, the subsidy might be limited to wells drilled at a distance greater than one mile from a producing well. The subsidy might also be structured so that it varied with the size of the leasehold on which the exploratory well was drilled. For example, the full subsidy might be allowed on a leasehold covering either full or partial interest in as much as 20,000 acres and be reduced proportionately for leasehold sizes in excess of 20,000 acres.

A further dimension of the information problem is related to the fact that property law in this country recognizes private interest in underground oil and gas deposits. Even if the information problem could be readily solved by developing techniques for identifying underground oil and gas deposits (including federal and state governments), the result would be that landowners would have a greater opportunity to extract royalties. The royalties, however, include a windfall element. They could be more heavily taxed, as Stiglitz shows, without appreciably changing the pattern of oil and gas development or production. This line of analysis suggests that special attention must be given to income from non-operating interests, that is, royalties. We will deal with this in Chapter Four.

NOTES TO CHAPTER THREE

[1]The qualifications of this statement do not change its basic thrust. If there is unemployment, we can increase production, but there remains the choice of whether to increase output of energy or of something else. Over time, we can increase total production, but again there is the choice of whether there should be more of an increase in the form of energy output than existing markets call for, or whether more resources should be devoted to increasing other outputs.

[2]This analysis leans heavily on Joseph E. Stiglitz, "The Efficiency of Market Prices in Long Run Allocations in the Oil Industry," and Frederick Peterson, "Two Externalities in Oil Exploration," in *Studies*.

[3]Senate Interior and Insular Affairs Committee, Senator Jackson, Chairman, *Hearings on Financial Requirements of the U.S. Energy Industry*, March 6, 1973.

[4]*Ibid.*, p. 54.

[5]If capital markets are badly organized, this hurts industries other than the energy industries, and one must deal with the problem in a capital market study.

[6]S. McDonald, *Federal Tax Treatment of Income From Oil and Gas*, op. cit., and S. McDonald, "Distinctive Tax Treatment of Income From Oil and Gas Production," *Natural Resources Journal*, 10, January 1970, pp. 97–112. There was an extensive debate on this in volumes 15 to 17 of the *National Tax Journal*, with critiques of the McDonald argument by D. Eldridge, R. Musgrave and P. Steiner, and rebuttal by McDonald.

[7]We include in the loose term "borrowers" firms selling new issues of common stock.

[8]S. McDonald, "Distinctive Tax Treatment of Income from Oil and Gas Production," *op. cit.* McDonald concludes from a careful restatement of this argument that at best it justifies a percentage depletion deduction of 14.5 percent instead of 22 percent.

[9]When the profit is 15 percent of final price, the tax would be half of this, but adding the tax to the price increases the amount taxed. When the price has increased by 15 percent, then the full tax has been shifted and the taxpayer is left with the same 15 percent rate of return he would have had with no tax.

[10]The economic literature on whether the corporate tax is, in fact, shifted into higher prices is extensive, and since our conclusion does not depend critically on our no-shift conclusion, we will deal with it briefly: in the short run, the corporate tax should not affect price because if the business could have earned a larger before-tax profit by raising price, it would have done so even without the tax. This simple argument could break down in some empirical testing. The work of Kryzaniak and Musgrave, *The Incidence of the Corporation Income Tax,* Baltimore, Johns Hopkins Press, 1963, suggests that the tax is shifted into higher prices. We side, however, with the numerous critics of Kryzaniak and Musgrave (see particularly R. Gordon, "The Incidence of the Corporation Income Tax in U.S. Manufacturing," *American Economic Review*, 57, 1967, p. 732). In the long run, the corporate income tax could be shifted into higher prices if it reduces the return on capital and thereby reduces the supply of capital. The problem in

this sequence is that there is much evidence that the supply of capital (savings) is not much affected by the rate of return. See Cragg, Harberger, and Mieskowski, "Empirical Evidence of the Incidence of the Corporation Income Tax," *Journal of Political Economy*, 75, 1967, p. 811.

[11]The argument over whether the corporate tax works to the disadvantage of the oil and gas industry is explored and rejected in the first part of Professor Stiglitz's article, "The Efficiency of Market Prices in of Long Run Allocations in the Oil Industry," *op. cit.* He goes beyond the arguments that we summarize here by looking at the problem from the point of view of some recent economic literature on the theory of optimal taxation, which assumes that any tax will be "unneutral" in some ways and that the best tax is one that involves unneutralities that will not cause many changes. He concludes that responses to price changes in the oil industry are not very great (that is, are inelastic); so, based on the theory of optimal taxation, there would not be a bad result even if the corporate tax resulted in a higher price for oil.

Stiglitz's result depends primarily on the reasonable estimate that the elasticity of supply in the oil industry must be lower than the corresponding long-run elasticity of supply in manufacturing industries. Essentially, manufacturing facilities are reproducible in the long run without much likelihood of rising costs. Natural resources, however, are more likely to involve increasing costs as the most accessible reserves are used up. On these grounds the optimal taxation argument for a higher tax on oil companies is substantially the same as Pigou's argument for a tax on increasing cost industries. See A.C. Pigou, *The Economics of Welfare*, 4th edition, London, MacMillan, 1952.

[12]John McLean, Continental Oil Company, Testimony before the Ways and Means Committee, U.S. Congress, *General Tax Reform*, February 26, 1973, pp. 1,253-1,254; Richard Gonzales, *ibid.*, pp. 1,349-1,354.

[13]*Petroleum Facts and Figures, 1971*, p. 31.

[14]Cf. C. Jackson Grayson, Jr., *Decisions Under Uncertainty: Drilling Decisions by Oil and Gas Operators*, Boston, Research Division, Harvard Business School, 1961.

[15]The natural gas part of the oil and gas industry has some protection against this kind of risk due to the prevalence of long-term contracts.

[16]S. McDonald, *Petroleum Conservation in the United States: An Economic Analysis, op. cit.*, p. 38.

[17]It does not change the argument if we require mine owners to use safety equipment, etc., to reduce risk in the same way that we have conservation regulations to reduce risk in the oil business. The point is that the net-risk costs will be reflected in the price of coal and should not be offset by government subsidies or tax concessions.

[18]See "International Aspects of Energy Policy," Brookings Institution in the Energy Policy Project series.

[19]Professor Peterson discusses this, *op. cit.*

[20]The economists' term is to internalize them, that is, bring them inside the drillers' decision-making process.

[21]Other such arrangements are described in Grayson, *Decisions Under Uncertainty: Drilling Decisions by Oil and Gas Operators, op. cit.*

[22]Edward Miller, "Some Implications of Land Ownership Patterns for Petroleum Policy," *Land Economics*, 46, November 1973, pp. 413-423.

[23]Peterson, *op. cit.*

Chapter Four
Royalties and Non-Replaceable Resources

4.1 ROYALTY INCOME

Before turning to specific policies, it is necessary to discuss another feature of the economics of natural resources: A free market for natural resources will give rise to a peculiar kind of income called royalty income.[1]

When producers drill for oil and gas, an essential requirement is legal title to drill and ownership of at least part of what comes out of the well. Legal title is generally acquired through royalty contracts, which give the landowner a specified share of the value of the product, say, 15 percent, and by lease bonuses, which are lump-sum payments to the landowner.

It is important to talk about royalty income from natural resources separately because a royalty, such as land rent, is essentially a surplus income. The landowner need not do anything to earn this income except divert some of his land from other uses. The sale of some mineral rights can command a price that bears little relation to the value the land would have had if used for an alternative purpose. For land that previously had a particular use, such as cattle grazing, there may not even be much need for diversion. Instead, the value of the mineral rights relates to the scarcity of good mineral rights of that kind.[2]

The legal systems of property ownership in the United States recognize titles to subsurface minerals, which accompanies ownership of surface land. This system is unique among the major oil-producing countries, where subsurface mineral titles usually belong to the state.[3]

As one might expect, the market value of subsurface mineral rights behaves very much like land rent. Any piece of land that has both

61

desirable and scarce characteristics will command a rent that is not related to reproduction cost, because most of the desirable characteristics are not freely reproducible. For this reason, the income earned from owning land is different from the income earned by owning a machine.

In the long run, a machine is reproducible. Thus, if a particular kind of machine can produce things that are desirable and scarce, its owner will command a large net income or profit. This income, however, has an economic function. It will cause other businesses to reproduce the machine, and so its products will become less scarce. The process of reproduction will go on until owning the machine yields barely enough income to cover the cost, plus a normal return on the capital invested.

An unusually high profit associated with a machine tends to be gradually wiped out because the machine is reproducible. On the other hand, income associated with ownership of land that has good mineral prospects does not get wiped out prior to exhausting the mineral deposit because there is a limited amount of land with drilling prospects. Clear evidence of this is statistics that show that the average depth of an oil well has increased between 1968 and 1950 from 3,680 feet to 4,627 feet.[4] Over time, we use up the deposits near the surface.

It does not change our argument to recognize that from time to time there will be breakthroughs that suddenly reveal the existence of hitherto unknown deposits better than the prospects that have been drilled. The discovery of the East Texas fields was such an event, and the discovery of the Persian Gulf fields would have been such an event except for their distance from U.S. markets and U.S. oil import quotas at the time. East Texas did reduce oil prices and royalties. This kind of reaction is, however, historically rare; the usual pattern is for progressively poorer drilling prospects to prevail.[5]

The policy significance of what we have been saying about royalty income lies in the combination of scarcity and desirability. As we described in Chapter One, a price system serves to ration available products among various claimants and to stimulate production. Some of the higher price goes into a higher royalty, which does not stimulate production because the royalty is associated with the basic scarcity of drilling prospects.

It would follow, then, that to the extent that tax benefits act as price increases, some of the benefits will go into increased royalties and will not serve to increase production. Before exploring policy matters in detail, we must turn to the statistical problem of estimating the relationship between prices and royalties.

4.2 ROYALTIES AND SUPPLY ELASTICITIES

It is conceivable that (1) there is a great quantity of land with prospects for finding oil or gas and that (2) there is not much difference in the prospects between tracts. If this situation really exists little will be paid to the owner of any particular prospect because a firm wanting to drill can readily find as good a prospect. Also, if the demand for oil and gas rises, there will be little need to divert much of the price increase to land payments. Nearly all of it can be used to pay for more drilling equipment and for more oil and gas.

It is also conceivable that oil and gas drilling prospects are highly individualized, that some are much better than others. To simplify the complex ways in which prospects may differ (location relative to markets, hardness of caprock, etc.) we can think of the difference as the average expenditure on drilling per barrel of oil obtained. If the average expenditure varies dramatically between drilling prospects, then an increase in price will require heavy equipment expenditures on additional wells.

The chart below makes a hypothetical comparison of the two situations.

	Average Drilling Expenditure Per Barrel of New Reserves	
Quantity of new reserves (billions of barrels)	Case 1 Fairly uniform prospects	Case 2 Differentiated prospects
1	.75	.75
2	.77	1.00
3	.79	1.25

Consider a market in which there is a demand for an addition to known reserves of only one billion barrels. In either Case 1 or Case 2 this could be accomplished on properties for which the average drilling expenditure is 75 cents per barrel. Now consider how much price must increase to make it profitable to find two billion barrels. In Case 1 only slightly lower quality prospects will have to be explored. The price of oil must rise to cover the extra drilling expense, and the acres that would have been drilled even with the lower demand will be able to command an extra royalty of 2 cents per barrel. In Case 2 to get two billion new barrels the price must rise more sharply; and now the acres that would have been drilled anyway can command a royalty of 25 cents more.

The extent to which prospects are differentiated can be described by a concept called "elasticity of supply," first mentioned in Chapter Two. (In Case 1 of our example, the case of fairly uniform prospects, we would say that the supply is elastic.) Davidson, Falk, and Lee have explored the royalty issue by estimating elasticity of supply. They find that, under very plausible assumptions, one could expect the elasticity of supply of a natural resource to be equal related to the royalty in the following simple way:

$$\text{elasticity of supply} = \frac{1-A}{A}$$

A equals the percentage of royalty in relation to the value of the resource at the mine or well-mouth. This relation suggests that when supply is very elastic—the Case 1 situation—the royalty share will be low. When supply elasticity is *not* very elastic, the resource is very scarce and high price will not increase output—the Case 2 situation—and the royalty share is high. A royalty share of one-half is consistent with supply elasticity of 1.0 because the royalty share should be half of production revenues. The importance of determining elasticity of supply is that, from historical data on royalty shares from oil production, the authors are able to draw some conclusions about the way in which tax incentives for oil-well drilling may generate increased royalties. The significance of this is that from the standpoint of the driller, these increased royalties are a cost that tends to offset some of the incentive effect of the tax benefits.

4.3 ROYALTY INCOME AND PRODUCTION TAX BENEFITS

Such tax law provisions as percentage depletion or intangible drilling expenses serve to increase the prospective profit from an oil or gas find. The right to drill on land with some decent prospects that have been established by geological and geophysical studies is, so to speak, an essential condition for enjoying the tax benefits. This whole discussion of the royalty situation, however, is based on the point that there is not a great supply of land with good drilling prospects.

It follows from this that if government does things to make drilling more profitable, some of the benefit can be captured by the landowners. This transfer of some of the tax benefit into the hands of landowners would be a problem even if percentage depletion were not allowed on royalty income. Knowing that drilling rights on a piece of land are valuable makes it possible for the landowner to charge for drilling rights. If drilling rights are made even more valuable by enactment of percentage depletion, drillers would be willing to pay even more for these rights. Landowners get more for the rights by letting drillers bid against each other.

The Davidson, Falk, and Lee analysis suggests that the elasticity of supply for United States onshore drilling rights might be as high as 3, which in turn suggests that about one-quarter of the gross from drilling is captured by landowners through royalties and lease bonuses. This implies that one-quarter of the tax benefits for producers are dissipated into more royalties.

For offshore drilling (on the outer continental shelf) elasticity of supply is apparently lower, and perhaps as much as 40 percent of the tax incentives are captured by royalties. The fact that these royalties go to the federal government—and implicitly to the public—means that the tax incentives for offshore drilling are less generous than they would appear to be if we were looking only at the tax law.

4.4 ROYALTIES AND MONOPOLY

Another aspect of the royalty situation is that royalties are affected by the presence of monopoly elements as well as by the basic scarcity of drilling prospects in nature. Just as man has found ways to improve on some of nature's scarcities—by making rain, for example—he has found ways to make natural scarcity more profitable by superimposing monopolistic constraints.

Monopoly might operate in several ways. If the ownership of land with good oil and gas drilling prospects were highly fragmented, but there were only one drilling firm or a tightly knit combine of drilling firms, these firms would enjoy a monopoly profit. The monopoly could gradually divert a large part of increases in consumer demand into a higher price by providing only a limited increase in output.

Alternatively, monopoly power might reside with the owners of oil lands, which is the case with offshore drilling on the U.S. outer continental shelf and with oil and gas lands in the less-developed countries that have been organized into a cartel—the Organization of Petroleum Exporting Countries (OPEC).

In the case of the outer continental shelf, the United States government has followed a policy of being extremely restrictive in opening offshore tracts to competitive bids, which has kept up bid prices. The OPEC strategy has been to limit possible price competition between the relatively large firms that have drilling licenses in the less-developed countries by requiring a fixed payment per barrel to the host country. This strategy has not reduced the profits of the drillers much because the payment has been passed on as price increases to consuming countries.

In both cases the important kind of restraint on output appears to arise from monopolistic restriction imposed by the landowners. A powerful tool for understanding the operation of this restrictive policy is the concept of user cost.

From the landowner's point of view, if he sells the rights to drill on his property this year, he will not be able to sell them next year. If there is a strong prospect that oil and gas prices will rise, it would be a good tactic for the landowner to hold out for higher current payments for drilling rights to compensate himself for the loss of future income.[7]

This potential for change in the value of the landowner's wealth associated with changing oil and gas prices is called the user cost. When prices are rising, user costs are positive. There is a cost in selling drilling rights now rather than next year. When prices are expected to fall in the future, user costs are negative, and owners become eager to sell.

4.5 DOMESTIC POLICY IMPLICATIONS

In Chapter Twelve we will return to the problems of royalties relating to United States oil import policies. At this point, we limit ourselves to some policy implications for domestic energy policy.

From the standpoint of domestic policy, the present extension of percentage depletion to royalty income from U.S. onshore drilling is bad. This tax benefit does not increase oil and gas output because the royalty contains considerable windfall elements. The royalty contract must pay the landowner enough to compensate him for any income lost through diversion of the land from other uses, but in general the loss is likely to be quite small in relation to oil royalties. (This will not always be true. To cite an extreme example, if there were oil under Manhattan Island, the royalties would probably not be high enough to cause landowners to divert the land from other uses.)

A special circumstance affecting the economic behavior of landowners in the United States is the "rule of capture," which holds that "petroleum ultimately belongs to the landowner (or lessee) who captures it through wells located on his land, regardless of its original location as a natural deposit."[8] The effect of this rule is that a landowner is under considerable pressure to sell rights to oil or gas that might be "capturable" from his land. Because the common geological structure of an oil pool is a body of oil with gas or water drive underlying several legal tracts of land, the amount of ultimately recoverable oil can frequently be obtained by a well or wells driven on any one of several tracts. The rule of capture says that the oil belongs to the owner (or lessee) of the tract through which the oil is captured. A landowner does not own oil just because it is under his land; he also has to draw it out through h s land, and he has to do it before someone beats him to it.

The broad implication of this rule is that a considerable part of oil and gas policy in the United States has in fact been devoted to "slowing down" production from known reservoirs. This slowing down policy is called conservation, which is an appropriate name. Unrestrained drilling by landowners can lead to waste of drilling resources and a dissipation of natural pressures within the oil reservoir that might reduce the amount of oil ultimately recoverable.

Some of the specific conservation policies that have come into being are well-spacing requirements, limits on production from particular wells, and requirements that the various tracts in an oil field be operated together (or "unitized"). The thrust of all these policies, for our present purposes, is that there is even less need to worry about the economic incentive to a landowner to sell his oil. Even if the percentage depletion allowance is retained, it would be wise to disallow percentage depletion for royalty income.[9]

A sudden change in policy creates potential problems for investors who have recently purchased a property interest entitling them to a mineral royalty income, if the price paid for the property was higher in anticipation of a tax benefit. The present high price of oil, however, makes this problem trivial because royalty recipients will enjoy windfall benefits from the price rise.

If a rule were adopted requiring different treatment of depletion on royalty and non-royalty income (or between operating and non-operating interests), some special provision would have to be made for the operator who owns land outright. Essentially, an average royalty formula that assumes part of the income was royalty would have to be constructed.

As we point out in Chapter Ten, royalty income is remarkably concentrated in the higher income brackets, so that extending favorable tax treatment to royalties sharply reduces the progressive effect of the tax system. The situation is aggravated by the form of the tax incentive, which allows 22 percent of the income to be tax free. This makes the value of the tax benefit increase with the income of the recipient.

Another policy issue is the strategy of the government regarding leasing of offshore drilling rights for oil and gas. As a conservator of the public interest in these rights, the government should decide its current leasing policy with regard to its expectation of present and future prices of oil and gas. Due to the politics of federal budgetary policy, however, there will be some pressure to limit the amount of land put up for bids in order to maximize the current payments and reduce the budgetary deficit. In the present study we do not have sufficient information to set out a detailed strategy for lease policy, but we do express the judgment that it seems to be too restrictive. A more aggressive policy would strengthen the expectation of higher oil output in the decade of the 1980's and be a strong element in making oil production in the 1970's more attractive for the OPEC countries.

A dimension of this problem is an offshoot of the system of property titles in the United States and can be compared with the OPEC countries' ability to revise their "royalty" agreements by changing tax laws. With great uncertainty about the future of oil and gas prices, this kind of drilling contract involves unnecessary bureaucratic risk. Government would look foolish and perhaps dishonest if it should turn out that increasing oil and gas prices later proved that the government had made spectacularly bad deals. The appropriate procedure is to look for more flexibility in terms. One way to achieve this is to consider the application of a severance tax if royalty rates on new contracts rise appreciably.

NOTES TO CHAPTER FOUR

[1] Unless the context clearly specifies otherwise, we will use "royalty income" to refer to all the income of the landowner-royalties proper, lease payments, lease bonuses, etc.

[2] For there to be any royalty income, the value of the mineral rights must be at least greater than the value of income lost by whatever diversion from other uses is required to extract the mineral.

[3] M. Adelman, *The World Petroleum Market*, Resources for the Future, Baltimore, Johns Hopkins Press, 1972, p. 44.

[4] *Petroleum Facts and Figures, 1971,* pp. 32–34.

[5] M. Adelman, *op. cit.,* pp. 66–73.

[6] For a detailed derivation of this relationship see Davidson, Falk, and Lee, *op. cit.,* in *Studies.*

[7] The technical situation is different in the OPEC countries because the owners of the mineral rights—the governments involved—retain the right to change the royalty terms more or less at will by imposing tax and pricing rules on the operating companies.

[8] S. McDonald, *Petroleum Conservation in the U.S.: An Economic Analysis, op. cit.,* p. 31.

[9] In Chapter Five we argue it would be better policy for the United States to eliminate percentage depletion altogether.

Tax Benefits and Other Subsidies To Provide Additional Domestic Natural Resources for Energy

5.1 THE PROBLEM AND ITS APPROACHES

Our analysis in Chapter Three concluded that there were some defects in the market prices related to natural resources for energy. Some of these, such as the information aspects, have led to specific proposals related to bidding, geological and geophysical research, and the like. Another defect is the security problem that arises when imported oil can dramatically undersell domestic oil. We will turn to ways of dealing with the security problem shortly.

Except for these two problems, there is nothing about the business of producing natural resources for energy that suggests the need for tax incentives.[1] We think that it makes sense to rely on market prices to direct resources into energy production. A higher price rewards all producers of energy whether they use up valuable natural resources or expend manufacturing efforts on cheap resources. Letting the market price of energy provide the incentive avoids the results of the present law, demonstrated in Chapter Two which provides that energy produced from oil and gas is heavily subsidized by tax benefits and other energy sources get little or no subsidy.

In addition to treating various energy producers unequally, the approach of reducing energy prices through producer incentives serves to beguile consumers into using too much energy. This point can be brought home strikingly by observing that, as a tool for relieving an energy shortage, the country is helped as much by an invention that increases energy output. Producer incentives that lower the price of energy reduce the pay-off from any technological change that would economize on energy use.[2]

The argument for encouraging the production of domestic energy resources in the United States becomes more complicated, however, when we introduce the possibility that the world price of oil can be well below the domestic price. In that situation free market prices would lead to imported oil becoming a major source of domestic energy. The domestic price would be driven down to near the world price and only those companies producing the lowest-cost domestic sources would find it profitable to operate.

The potential problems of high imports of crude and residual oil faced the United States in the mid-1950's. Our response was to impose quotas on oil imports to relieve the threat to national security that was implicit in dependence on foreign sources.

This was not a matter of *military* security, because the military share of oil and gas output could easily be met from U.S. sources. Instead, the problem was the potential interruption of civilian uses as a result of hostile action against ocean-going tankers or restrictions on production by a number of producer countries acting in concert.

The tax benefits for natural resources antedated by nearly forty years the security problem of the 1950's. Nevertheless, when the security issue arose it gave powerful support to the tax benefits. Public hearings before the Ways and Means Committee in 1963 came at a time when the Committee was particularly sensitive to the tax reform issue, which had been elaborated in a compendium published by that Committee in 1959.[3] A leading witness for the oil industry in 1963 was General Lucius Clay, whose argument was concerned with the security implications of tax benefits to the domestic oil industry.[4]

Another dimension of the tax benefit-security connection was that the oil import quota imposed directly on domestic consumers of energy a cost that was related to the difference between world prices and domestic prices. Domestic consumers paid the higher domestic price because the quota was implemented by the governments distributing to refiners "free tickets" to buy cheap imported oil for resale in the protected market. The annual cost of this extra domestic price was $4 to 6 billion dollars.[5] Under these circumstances a tax benefit, which in part reduces domestic oil and gas prices, serves to relieve the protest against limitation on imports.

The background of these protectionist policies changed dramatically in 1973, when the international cartel, OPEC, acted to drive the world price far above the prevailing U.S. domestic price. Very rapidly the United States price began to climb toward the world price. By early 1974 oil companies were suddenly making such large profits that the President proposed a windfall-profits tax on sales of domestic crude, and the Senate was mumbling about massive rollbacks in prices.

The future is hard to predict. If the OPEC price stays high, there should be vigorous expansion of United States energy sources and, in five years or so, energy self-sufficiency for the U.S. If the OPEC price comes down, we can again have high oil imports and a renewed security problem.

Sections 2,3,4, and 5 of this chapter are devoted to an analysis of government policy toward energy resources, especially oil and gas, under the assumption that we will have a security problem if the world price is below the U.S. price. It is in this situation that we have had most of our recent experience with tax benefits, and experience is an excellent teacher. It is possible that we will face the situation again. Section 5.6 also examines the role of tax policy toward the producers of energy resources when world prices are above United States prices and the domestic industry is booming.

5.2   ALTERNATIVE ROUTES TO ENERGY SECURITY

United States energy supplies can be made more secure by three different developments. One is an increase in the share of our energy resources supplied by domestic producers. The smaller the import share, the more trivial its interruption. It is important to recognize, however, that this is a short-run response because increasing production from domestic sources involves a faster depletion of the best (cheapest) prospects and a continuous necessity to develop less promising prospects.[6] To the extent that government policy is directed at a greater reliance on domestic energy resources, including uranium and low sulfur coal, the higher the future costs of increments to output from these resources will be and the more difficult it will be for them to meet the competition of foreign prices in the long run.

A second development that could help solve the security problem is maintenance of extra productive capacity by U.S. domestic producers so that the U.S.-producer share of the market could quickly expand to offset import interruption.

The third development is maintenance of an inventory of crude oil that could be drawn on in the face of a supply interruption.

These three developments are really theoretical descriptions of possible program results or targets. We are interested in specific policies that the government could adopt, some of which might involve more than one result. The three developments are results to look for while we are listing the more specific operational programs open to government.

One kind of operational program is to enact tax measures that increase the profitability of producing energy from natural resources in the United States. This, of course, has been done.

Such a program can have several immediate price effects that will lead to one or another of our target developments:

1. The tax benefits can increase output and lower product prices.
2. The tax benefits can, in the short run, cause producers to carry more reserve capacity; but, in the long run, by inducing faster use of good reserves, they can reduce the U.S. reserve capacity.
3. The tax benefits can increase royalty payments.
4. The particular structure of tax benefits can lead to changing the production technology in ways that increase the cost of production.

It is clear that tax benefits under the present law have some of each kind of effect.[8]

A second kind of operational program would be to extend the tax benefits now given to oil and gas companies to other energy resources. This means more than just applying the present percentage depletion and deduction of intangible drilling expenses to other resources because these tax preferences provide relatively lower benefits on resources other than oil and gas.[9] A producer incentive program that has a more nearly neutral effect on all energy resources would involve creating new kinds of benefits. One might be to extend something similar to percentage depletion to the manufacturing costs involved in producing a more usable fuel from coal or shale, another might be massive research subsidies for technology and new fuel sources. There are some research subsidies today, particulary for nuclear generation and coal treatment.

A third kind of operational program would be to impose a tariff on oil imports to raise their price and implicitly allow a higher price, or a higher producer incentive, for domestic resources. From the standpoint of long-range concern with energy shortages and rational allocation of those resources, the tariff has the advantage of raising energy prices to consumers and discouraging consumption rather than lowering prices and encouraging consumption—which is what occurs under a producer subsidy. A tariff that imposes a price penalty on imports would directly increase the share of the U.S. energy markets served by domestic resources; and increasing the profitability of the domestic resource business should lead, in the short run, to increased productive capacity relative to output, that is, some reserve capacity.

A fourth kind of operational program is to provide extra capacity to produce crude oil that could quickly be put into operation. One way of effecting this would be to find a way to induce or require private firms to carry a reserve capacity greater than the minimum necessary for current production. A model might be the Elk Hills reserve, which is a government-owned oil field more or less ready for production that is being withheld in case it is needed. The government could provide additional reserve capacity in parts of the outer continental shelf.

A fifth kind of operational procedure is the outright purchase of crude oil by government to be held as inventory. There are some natural formations, such as salt domes or abandoned mines, or special facilities, such as metal tanks, where storage of large amounts of crude oil seems feasible. An inventory provides a reserve against future needs; in addition, the purchase of the inventory serves to increase resource prices and provide some producer incentive. The inventory-purchase programs of the Commodity Credit Corporation in the agricultural field, for instance, were specifically designed to operate as price supports during the period in which the CCC was building its inventory.

5.3 THE BEST SECURITY PROGRAM

The five operational programs we have outlined are not easy to compare because they lead to complex developments—a combination of more domestic production and more reserves for future needs. We can, however, say something about the alternative approaches on general theoretical grounds before we select the best approach.

The idea of building protection against future supply interruptions is a unique cost of the decision to use a resource that relies on foreign, that is, interruptible, supply. This suggests that the protection should not be free to the beneficiaries. To illustrate the economic point of this assertion, assume the following set of facts:

1. Electricity can be generated from oil at a slightly lower cost per kwh than coal.
2. Electric plants designed for oil operations cannot be readily converted to coal firing.
3. The security problem associated with oil imports requires some cost to assure supply security; this cost will be larger than the saving from using oil rather than coal.

Given these assumptions, it is inefficient to use oil for generating electricity because it involves two costs, the cost to producers and the cost incurred in dealing with the security problem. If government paid for the costs of security out of general revenues, it would be making an inefficient oil subsidy. If it dealt with the security problem by enacting producer subsidies so that more oil would be produced domestically, the distortion would be the same; the economy would be distorted into making the wrong fuel choice.

The example is unrealistic only in its assumption that *all* electric companies find almost the same differential between coal and oil firing. The cost comparison depends on location or, more specifically, on the delivered price, and so competition does occur in many cases. Many electric companies hedge between using coal and using oil. Other users besides electric companies have a choice between coal and oil, at least in locations where delivered prices are close. In the long run, if research on liquefication or gasification of coal is successful, there will be other ways for coal to compete with oil and gas. The same kind of competition arises between oil and gas and other fuels, such as uranium, oil from shale, and solar energy.

It follows from this analysis that it would be a mistake to deal with the oil security issue by making domestic oil cheaper. The rational procedure is to recognize that the protection from possible fuel interruption is a special service being rendered to users of oil, and that the price of oil should reflect this circumstance. Even with the cost reflected in the price, most oil and gas consumers would continue to use those fuels. But some, such as automobile drivers, might consume less, and others would shift to other fuels. It is fair to make oil and gas users pay for the security of service they get, and if they decide to use less oil and gas (because some of the oil and gas is not worth the price of the product plus the price of the security), a better allocation of resources will result.

That the cost of customer security should be paid by customers constitutes a powerful argument against the first two operational programs listed above: namely, existing tax benefits for the production of energy minerals and some equivalent extension of these benefits to shale, coal, and uranium.

Percentage depletion should be repealed for all.[10] We also think that the deduction for intangible drilling expenses on successful development wells should be terminated. As we saw in Chapter Three, a plausible case can be made for externalities in exploratory drilling, but in fact the current practice of deducting dry-hole costs assures that the bulk of capital investment in exploratory drilling will be treated better from a tax standpoint than normal business investment. Consequently, a program of complete repeal of the deduction for intangible drilling expenses on successful wells comes reasonably close to a neutral tax policy, that is, one that does not put the cost of customer security for fuel users on the federal government.

Before commenting further on the effect of repealing these

tax benefits, we want to indicate that the security argument does lead to some other operational programs that would replace, in part, some of the incentives involved in the tax programs.

Our third, fourth, and fifth operational programs can be reconciled with the principle of putting the cost of security on customers. The third one, import controls (either quotas or tariffs), is in itself only a short-run response to the security problem. Basically, import controls cut down the portion of the market supplied by foreigners and increase the portion supplied by domestic sources. This necessarily reduces the *long-run* capacity to produce from U.S. sources.

The long-run damage to U.S. capacity from using our oil faster can easily be exaggerated into "we will use up all of our oil," but the effect is none the less real. The 1969 estimate of original oil in place (before any drilling started) in all U.S. fields discovered through 1959 was about 372 billion barrels. As of 1959 the cumulative production from these fields was 60 billion barrels; 32 billion barrels of the remainder were classified as "reserves," that is, they were reasonably recoverable from known reservoirs under existing economic and operating conditions. In total, 25 percent of the known oil in place had been either recovered or was imminently recoverable. Ten years later, because of new discoveries, the estimate of original oil in place in known U.S. fields was up to 395 billion barrels. The cumulative production was then 90 billion barrels and reserves were 30 billion; so 30 percent of the known oil had been either recovered or was imminently recoverable.[11]

If we estimate that 50 percent of original oil in place will be ultimately recoverable, the production between 1959 and 1969, along with the discoveries in that period, reduced the potential future production from U.S. and Canadian sources from 126 billion barrels to 108 billion barrels.[12] Current production is around four billion barrels a year, and the complete elimination of imports could raise it toward five billion barrels. In the future we will have new discoveries; the Alaskan finds, for example, have added about 10 billion barrels to reserves. It is clear, however, that a decision to reduce imports substantially does bite quite heavily into U.S. productive capacity over a decade or two.

The most promising ways of achieving a secure oil supply come down to our fourth and fifth operational programs, which involve a combination of imports and specific reserve provisions. They could take four explicit forms:

1. Establishment of a program to maintain a government reserve inventory of crude oil.
2. A requirement that refineries maintain specified levels of inventory of crude oil with or without payment for storage costs.

3. Acquisition of government-owned shut-in oil reserves (that is, fields that have been explored with the existence of reserves established but with no oil being produced).
4. An arrangement to require or induce private producers to maintain higher levels of reserves.

All of these programs involve costs: inventory programs involve storage costs; keeping shut-in reserve capacity involves maintenance costs; both programs involve a lost return on capital that is tied up in inventory or reserves. The costs should be met by a tax on imports. The effect of the tax would be to raise the price of imported oil, which would improve the market for domestic producers. As in our earlier example, fuel costs depend on location. There is no single point at which all U.S. oil markets go foreign or stay domestic. When foreign prices are low, they can capture markets near East Coast ports as well as inland. As the landed price of imported oil becomes higher, imported oil can only be sold nearer the ports. There are also chemical differences in crude oils that cause different refineries to have different critical price differentials that would lead the refineries to shift between foreign and domestic. Thus, any increase in foreign oil prices increases the market for domestic oil or oil substitutes, such as domestic coal, uranium, or shale.

5.4 SPECIFIC RESERVE STRATEGIES

The policies we have suggested for dealing with the oil security problem involve either reserves of oil in natural deposits (*in situ*) or inventories of oil extracted from natural deposits. A choice between these two depends on whether we are concerned about a short-run or long-run interruption of foreign supply. The geology of oil-well production is such that there is considerable loss in the amount of oil ultimately recoverable if the annual rate of production moves much above 10 or 11 percent of the estimated reserve.[13] Further, there is about a 60-day wait before production from a shut-in well can start. As it turns out, if the security "crisis" were to last as long as two years, natural reserves of oil are the least expensive technique for dealing with it. If the security crisis amounts to, say, a six-month or one-year interruption, an inventory of crude oil is the least expensive technique. We have not tried to evaluate the political dimensions of the security risk, so we have assumed that long-term and/or short-term extra capacity are needed.

We already have a long-term reserve capacity in the form of government-owned Naval Petroleum Reserves (NPR).[14] The one viable NPR at present is that at Elk Hills, Cal., which is estimated to contain about 1.4 billion barrels and could go into production to supply about 4 percent of the current U.S. output. The other large NPR is in the North Slope of Alaska. Although reserves there are variously estimated at 4 to 20 times the Elk Hills reserves, special long-distance pipeline facilities would need to be constructed to make this a viable alternative.

A problem with *in situ* reserves, in addition to pipeline connection, is the possibility that oil can be drained from the reserve by production on adjacent land, which requires the government to go into production through offset wells to avoid the loss of its reserves. This can be avoided by purchasing oil rights in large property blocks. The annual cost to the federal government of such a program would be the sum of the interest cost on the purchase price, including the cost of initial development and construction of a pipeline connection, plus the annual property maintenance cost. Based on the Elk Hills experience, Mead estimates that the cost could be in the vicinity of 10 cents per barrel of reserves. If all of this cost were allocated to production in the first year after the security crisis, the total would amount to about $1.20 a barrel of oil produced. If it were allocated to production within two years, the total would be reduced to 60 cents a barrel.[15]

Instead of following normal commercial practice, the Elk Hills reserves could be developed on the basis of all-out production in the first year, a level almost three times the commercial production rate. This would require additional investment and some sacrifice of ultimate recovery. The annual carrying cost of this program spread over the first year of production would be about 40 cents a barrel.[16]

The alternative of paying private companies to carry excess reserve capacity was actually accomplished in the 1960's by a combination of oil import controls, tax benefits, and state prorationing. During that time the reserve-to-production ratio was around 12:1.[17] On the general assumption that an extraction rate of 10 percent is reasonably efficient, the 12:1 reserve level corresponded to a potential for expansion of production of 20 percent.[18]

It would appear to be quite expensive for the federal government to induce oil companies to carry "unused" reserves. For it to be profitable, the rate of return after tax on oil production investment must be in the neighborhood of 15 percent.[19] The carrying cost of excess reserves is far cheaper with government ownership, especially in view of the necessity for a supervisory structure to police private production rates and reserve estimates. (Policing reserves would also be difficult.) One way companies could be compelled to carry reserves, however, is if the government specified an additional tax on petroleum production, which would be reduced if the reserve-to-production ratio exceeded 10:1 and removed completely if it exceeded 12:1. On the other hand, programs that require companies to carry extra reserves ultimately increase the cost of domestic oil relative to imports and thereby aggravate the short-run security problem.

To meet the unsatisfied oil demand in the face of a six-month interruption of foreign supplies, the ideal solution is to maintain inventories of crude oil at locations convenient to refineries. As mentioned before, these inventories can be maintained in steel tanks (above ground) or in suitable natural depostis such as abandoned mines, salt mines, or cavities produced underground by nuclear explosions. The various techniques for holding reserves have problems that appear to be manageable by appropriate site selection. The cost of an inventory reserve to the federal government amounts to interest on the acquisition cost of the oil, and—as appropriate—interest, amortization and maintenance of the storage facility. Mead has estimated the annual cost of an inventory policy as follows:[20]

	quantity	cost per barrel of oil reserve
salt cavern	1 million barrels	$0.72
	2	0.54
steel tanks	10	0.74
abandoned mine	10	0.31
nuclear cavity	1	0.72-0.57

Because the security problem arises from import interruptions, it would be appropriate to put the cost of keeping reserves on imports, which could be done be imposing a tariff. The exact level should be determined on the basis of a final technical evaluation of the reserve alternatives, plus political judgment about the size of reserve needed. The arithmetic works this way: if we maintain a reserve of one full year of oil imports, the tariff rate per barrel of oil should be the same as the annual cost of maintaining a barrel of oil in reserve, which should be 75 cents to $1 per barrel.

The possible security problem we have been discussing is directed at future policy. Under present circumstances of high world prices and shortages in the U.S., it is impractical to accumulate inventory. The ideal posture for the United States would be to stand ready to accept imports, subject to a tariff, whenever they become available at the United States price level, and to buy for inventory at that time. If the inventory were flexible the United States would be in a position to reward those who cut price, which would tend to weaken the cartel.

A commitment now to protect an expanded U.S. energy industry, that is, enactment of a tariff if foreign oil again becomes low-cost competition, would probably be of some value in encouraging the expansion required to reach self-sufficiencey.

5.5 A FURTHER EVALUATION OF EXISTING TAX BENEFITS IN RELATION TO THE SECURITY PROBLEM

In view of the long history of endorsement of benefits by Congress, it is appropriate to examine the relationship of benefits to oil and gas production more closely.

It is hopeless to reconstruct the argument on which Congress has based its conclusions that percentage depletion and the deduction of tangible drilling expenses are wise policies. The original decision to allow discovery-value depletion as a deduction was made in 1918. It was made partly in response to intra-industry competition problems arising from the fact that some wells could use as the basis for depletion the value as of March 1, 1913—the date when the income tax started.[21] The adoption of percentage depletion in 1926 dealt only with a formula approximation of discovery-value depletion.

The committee reports for the 1918 income tax act contain no analysis of discovery-value depletion as an economically efficient incentive. The Senate Finance Committee report in 1918 contains a charming account of a prospector who has looked for years for a strike and is then taxed unfairly when it all comes in in one year. There may have been such a case in 1918, but it is hardly a description of the industry today. Surveying the situation in 1959, Professor Charles Galvin, in an essay not unsympathetic to oil and gas tax benefits, said, "The considerable argument on the subject has thus far not produced a definitive articulation of criteria on the basis of which a workable and acceptable tax policy could be developed."[22]

In the absence of an official theory we must construct our own. In Chapter Two we presented a theory of the effect of percentage depletion that assumed that the industry does not operate by carrying a larger stock of reserves relative to output when tax benefits reduce the cost of capital. The Chapter Two analysis assumed that, on the contrary, investment in more drilling would cause more production.

An alternative theory of the effect of tax benefits was that tested by the CONSAD Corporation under contract with the Treasury Department.[23] Specifically, CONSAD assumed that the price was fixed, which in turn fixes the production; the only industry response to tax benefits would be to become more capital intensive, that is, to carry a larger ratio of reserves to output. CONSAD also found little evidence of a larger reserve ratio.

For convenience we will call the two theories the output expansion theory (the theory followed in Chapter Two) and the reserve expansion theory (the one explored by CONSAD).

If we accept the output expansion theory, the effect of percentage depletion and the deduction of intangible drilling expenses is a reduction of the price of oil by 10 percent and an increase of output by 3 percent.

In the first section of this chapter we related output expansion to the security problem and found this approach wanting. We argued that price reduction and production increase make minimal contributions to solving the security problem because the result does not change the level of imports much. The output expansion approach is also bad because it provides unequal benefits for different energy forms that could potentially compete with imported oil. Further, it is indefensible because increasing U.S. output to reduce imports will entail higher U.S. cost in the future.

While a pure tax benefit strategy that works by increasing domestic production and reducing price is an inadequate response to the security problem, it is significant that the proposed import tax will generate the desired effect as a sideline. An import tax of 75 cents per barrel would produce a considerable price increase for U.S. domestic oil whenever the world price is low enough to threaten the U.S. oil industry.[24]

This is an important point. A permanent feature of the U.S. tax law should not be offered as a defense against an import problem that may exist one year but not the next. If we are sincere about wanting to deal with the problems caused by excessive imports, we need a policy that is keyed to the *proportion* of imports. A program of inventory reserves and an import tax would operate this way.

The appropriate response to a security problem is to take advantage of whatever low foreign price develops and to have inventory reserves to deal with interruptions. This response is not only keyed to the problem, but it is self-adjusting. If OPEC insists on such a high price that it prices itself out of the U.S. market, then imports will decline and production from U.S. sources will rise. The decline in imports would mean lower collections of import tax; it would also mean lower levels of inventory reserves because, according to our proposal, the inventory level was to be related to annual import levels.

Dealing with an import problem by subsidizing U.S. production through tax benefits would, in principle, call for continuous revision of the tax benefits depending on year-to-year changes in the level of foreign versus U.S. prices. There is no automatic procedure for reducing the tax benefits if OPEC does, in fact, price itself out of the U.S. market.

Because the output expansion theory of tax-benefit effect does not generate a case for tax incentives comparable to the strategic inventory plan, we will turn to an investigation of the reserve expansion theory. Through most of the 1950's and 1960's the reserve expansion theory was probably operating because of a peculiar combination of historical circumstances. Specifically, a system of prorationing and import quotas was available to restrict output and prevent price reduction. This explanation was quite explicitly stated by General Ernest Thompson of the Texas Railroad Commission, which ran the state prorationing system:

> Chairman (Wolverton): . . . That kind of practice [setting allowables below what could be properly produced under the best engineering skill] . . . has the effect of increasing price to the consumer, is it right?
> Mr. Thompson: Yes, sir, it is.
> Chairman: On what basis?
> Mr. Thompson: Because you cannot ever build up a reserve supply for defense of this country unless you have incentive to build up this reserve, and it must be carried in the price of the product.[25]

Our problem is to separate the increase of reserves that can be attributed to tax incentives and the increase that can be attributed to prorationing. There are at least three strong pieces of evidence that tax incentives do not increase reserves.

Professor Kahn, in a careful review of the operation of percentage depletion against a background of state prorationing systems, concluded that oil companies had indeed increased their productive capacity relative to their rate of production, but that during the 1950's and early 1960's the share of exploratory wells in total drilling fell off. The increase in productive capacity had been in development wells, a relatively low-risk operation. Under a prorationing system in which more-productive wells are limited in the number of days a month they can operate, an operator of a good oil field can increase his allowable production by drilling more wells in the same oil deposit.[26]

The thrust of the experience under state production controls combined with tax incentives is that incentives created for increased capacity arose from production control. Among other effects, production control, which affected only the more productive wells, increased the cost of production.

A more basic analysis of the production theory of oil and gas wells is developed by Professors Wright and Cox.[27] They have set up a full econometric model of the oil and gas producing industry that involves treating the investment in reserves as a capital stock. The neoclassical theory of investment implies that measures that reduce the cost of holding capital stock, or that increase the profits attainable from holding capital stock, will lead firms to increase their investment in capital stock in a predictable way.

This is a widely held economic theory and it has proved fruitful in examining investment tax credit; a number of writers have found evidence that in the manufacturing industry, as a result of a better rate of return, investment in capital increases relative to output.[28]

To explore the reserve increasing effect, Wright and Cox investigated data covering the period 1959 to 1971, including annual reserves, drilling, capital costs, output, prices, and the level of allowable production in states with prorationing.[29] They analyzed oil and gas data separately as well as together, and they explored behavior in both prorationing states and nonprorationing states.

The Wright-Cox analysis does suggest that a deliberate policy of preventing the operation of some wells at full capacity could explain the presence of reserves redundant in relation to current output. On the other hand, a producer incentive—either in the form of lower reserve-finding costs (due to the deduction of intangibles) or in the form of a better effective price (due to percentage depletion)—does not increase reserves relative to output, that is, these tax measures do not produce a cushion of unused reserves that could be called upon in the event of an interruption of foreign supplies. All they do is reduce production cost, and to the extent that tax reduction does not go into higher royalties it increases consumption through somewhat lower prices. Increased consumption, of course, entails a demand for higher reserves—though these reserves are not a cushion to meet future interruptions, but merely part of the "civilian demand" now regarded as vital "needs" that must be filled in one way or another.

A brief statement of the Wright-Cox results in mathematical form are appended to this chapter.

The Wright-Cox conclusion does not differ substantially from that of the CONSAD study, which examined prior drilling, reserve, and price data for evidence on how, if prices and production are fixed, additional profitability after tax would be reflected in reserve levels. The CONSAD study found very little reserve effect.[30] Kahn also concluded that extra reserve carried in the past was not the result of tax benefits but the result of state prorationing.

It is interesting that in its critique of the CONSAD paper, the American Petroleum Institute made the same point that we are making here, that tax benefits do not change the reserve-to-output ratio: "...the required level of reserves is technologically determined by the level of production.... To produce one barrel of oil annually there must be about 10 barrels of reserves in the ground."[31]

Clearly, the assertions of the American Petroleum Institute cannot be taken at face value, because during much of the 1960's the ratio between reserves and production was not 10:1 but 12:1. As we have said, careful analysis of the reserve capacity during that time leads us to believe that the higher reserve rate was only the result of prorationing.

As a technique for providing national security, tax benefits are ineffectual. All that can be expected of tax benefits for production of a natural resource, such as crude oil, is a reduction in price, a small increase in production and a proportionate increase in drilling to keep reserves in line with production, and only a negligible increase in the reserve-to-output ratio. Because of tax benefits, we are slightly less dependent on imports, and domestic production might be 82.4 percent of consumption rather than 80 percent. (The extra 2.4 is 3 percent of 80 percent. The 3 percent effect was established in Chapter Two.) The extra production, however, does not provide a margin of reserve to deal with interruptions.

5.6 TAX BENEFITS WHEN UNITED STATES PRICES ARE BELOW FOREIGN PRICES

In 1974 oil taxes were again the subject of Congressional attention. As a result of the enormous increase in world prices of crude oil in late 1973, the President was calling for imposition of a windfall-profits tax—though, strangely, he was not calling for repeal of protectionist tax benefits.

Because the crude oil price in the United States rose from $3.50 per barrel in early 1973 to about $7 in early 1974, and there are prospects for an even higher price, the term windfall is appropriate. The Administration proposal would impose an excise tax on crude oil sales at prices above $4.50 according to the following schedule:

Excess price	Bracket rate	Tax at top of bracket
$ .01-.25	10%	$.025
.26-.60	20	.095
.61-1.20	30	.275
1.21-2.00	50	.675
2.00 up	85	

The schedule is to be modified by the zero tax level of price rising over the next three years until it reaches $7. By 1977 there would be no tax on crude sales up to $7. The tax would terminate in 1979, five years after enactment.

There is no satisfactory explanation in the Administration plan of why, in this period of booming oil sales, there should be less-than-normal income taxes on profits from selling crude oil at prices up to 100 percent higher than those of a year ago (which is the implication of an average price of $7 in 1974). Nor is there an explanation of why there should be a *combination* of less-than-regular-taxes on income plus a windfall excise tax on sales above a certain level.

The structure of the proposed windfall-profits tax is peculiar because it disregards costs; it is basically an excise tax. Consider the case of crude oil that might cost $6.25 to produce and that, in turn, could be sold at $6.50 a barrel. Under the President's proposal it would be taxed at 68 cents, which would cause the producer to lose money. Now consider the operation of percentage depletion in this case. For oil that costs $6.25 to produce and is sold for $6.50, the percentage depletion allowance is limited to 12½ cents—by the 50 percent of net income limitation. For oil that costs only $2.50 to produce, which would have been produced even at the old sales price of $3.50, the full 22 percent depletion allowance can be taken. The efficient well, which would have produced anyway, is getting windfall profits—not the marginal well. The two wells, however, pay the same excise tax; thus, the low-cost well gets almost 12 times as much tax-free income as the high-cost one.

The example of high- and low-cost wells makes it clear that the President's proposal completely misconstrues the concept of windfall profits. In economics "windfall" means being paid generously for what you were going to do anyway. The function of rising prices is to bring about new production, as demonstrated by crude oil price increases in late 1973 that brought about a flurry of activity to increase output.[32] Increased output, in turn, increases wealth in a society as long as a buyer is willing to pay more than the cost of producing a product. An excise tax that ignores cost will irrationally limit output in cases where there are no windfalls. The tax benefits, on the other hand, will continue to make a large part of the true windfalls tax-free.

As of late 1972 it was common to speak of long-run shortages in U.S. domestic energy output to forecast a large increase in oil production in the United States for 1980, when prices in the neighborhood of $4 to $5 per barrel were still being projected. This oil will still be produced even if the 1980 price is twice as high as previously expected, and oil companies and owners of oil royalties will, under the President's proposal, continue to pay less-than-regular business income taxes on this extra income after the windfall-profits tax has expired.

Another defect of the windfall-profits tax approach arises from the importance of price expectations in oil production economics. Crude oil cannot be "made," it can only be taken from a natural deposit. The oil can be withdrawn only once, and the production decision is essentially a matter of when to withdraw it. If the net-after-tax price of oil is expected to be substantially better in the future than it is now, the producer will favor future production over present production.[34] The introduction of a temporary tax is bound to affect many marginal decisions about when to drill, or when to apply secondary or tertiary recovery methods; inevitably, the decision will be made in favor of delayed production. This is clearly an undesirable result when there is an oil shortage accompanied by the sudden difficulty of importing.

The current oil industry argument in defense of the tax benefits talks little of foreign competition. The principal argument advanced now is that the industry needs capital for expansion.[35] Clearly, there are needs for capital to expand, but no evidence is offered on why ample capital would not flow into a highly profitable industry. A pattern of substantial self-financing for the oil industry, which has developed as a result of lower-than-average income taxes, is not imperative for the future. Nor is it necessary for the oil industry to own the entire energy business. Oil companies have bought controlling interests in many coal companies, and even entered businesses as diverse as nuclear energy, real estate, and the circus. It is understandable why the owners of these business empires would tell Congress that they should pay lower taxes than other companies—so that they can buy up even more of the energy business.

If Congress is convinced that one of the things that will lessen an energy crisis in the United States is more future investment, it would be more sensible to lower the corporate tax rate and let all kinds of companies expand into the energy business instead of "creating" extra capital for large oil companies so that they can outbid all rivals.

NOTES TO CHAPTER FIVE

[1]This is also the conclusion of Joseph Stiglitz whose work we relied on in developing the argument in Chapter Two. See J. Stiglitz, "The Efficiency of Market Prices in Long Run Allocations in the Oil Industry" in *Studies.*

[2]A striking example is the contrast between low gasoline prices and the proliferation of gas-guzzling automobiles in the United States and high gasoline prices and the predominance of economy cars in Europe.

[3]*Tax Revision Compendium,* Committee on Ways and Means, U.S. Congress, Washington, D.C., Government Printing Office, 1959.

[4]Hearings on the President's 1963 Tax Message, Committee on Ways and Means, U.S. Congress, March 26, 1963, p. 3,673.

[5]The annual U.S. consumption in the late 1960's was about four billion barrels, and the difference between U.S. and world prices was $1 to $1.50 a barrel.

[6]For a detailed discussion of the tendency toward increasing cost in oil and gas production, cf. M.A. Adelman, *op. cit.,* pp. 16–24; also S. McDonald, *Petroleum Conservation in the U.S.: An Economic Analysis, op. cit.,* pp. 91–92.

[7]It needs to be emphasized that tax laws do not produce oil. They change the prospects of earning income in various ways. Our estimates of how much these changes in income prospects affect the real world were given in Chapter Two.

[8]With regard to effects on production, see R. Spann, E. Erickson, and S. Millsaps, "Percentage Depletion and the Price and Output of Domestic Crude Oil," in *General Tax Reform,* Panel discussions before the Committee on Ways and Means, U.S. House of Representatives, February 26, 1973, pp. 1,309–1,328; CONSAD Research Corporation, "The Economic Factors Affecting the Level of Domestic Petroleum Reserves," *Tax Reform Studies and Proposals,* U.S. Treasury Department, published by the Committee on Ways and Means, 1969. With regard to the royalty problem see P. Davidson, "The Deduction Allowance Revisited," *Natural Resources Journal,* January 1970. With regard to cost increase effects see Susan Agria, "Special Tax Treatment of Mineral Industries," in *Taxation of Income from Capital,* Harberger and Bailey, Editors, Washington, D.C., Brookings, 1969.

[9]See Chapter Two, especially Table 2.1.

[10]We do not think that there are any special circumstances that would justify continuation of percentage depletion for non-energy minerals, but we have not addressed this problem explicitly in the present study.

[11]*Reserves of Crude Oil, Natural Gas Liquids and Natural Gas in the United States and Canada,* American Petroleum Institute, cited in *Petroleum Facts and Figures,* 1971, p. 115.

[12]Apparently, the ultimate recovery rate is quite uncertain. Lovejoy and Homan cite the figure 45 percent, estimated by Paul Torrey in *Evaluation of U.S. Oil Resources,* Oklahoma City, Okla., Interstate Oil Compact Commission, 1962, in Lovejoy and Homan, *Economic Aspects of Oil Conservation Regulation,* Resources for the Future, Baltimore, Johns Hopkins Press, 1967, p. 196. As we rely on more complete secondary recovery, the cost of oil rises sharply.

[13]Stuart Buckley, *Petroleum Conservation,* Dallas, American Institute of Mining and Metallurgical Engineering, 1951, pp. 151–152.

[14]The following is based on Walter Mead and Phillip Sorensen, "A National Defense Petroleum Reserve Alternative to Oil Import Quotas," *Land Economics,* 47, August 1971, p. 211; Walter Mead, "The Cost of Storing Oil," Testimony before the Committee on Interior and Insular Affairs, U.S. Senate, May 30, 1973; Cabinet Task Force on Oil Import Control, *The Oil Import Question,* Washington, D.C., 1970.

[15]Mead and Sorensen, *op. cit.*

[16]Mead and Sorensen estimated this in their 1971 article at 32 cents. Adjusting current crude-oil prices to be comparable with the figures above brings the price to the higher level.

[17]*Petroleum Facts and Figures,* 1971, p. 111.

[18]S. McDonald, *Petroleum Conservation in the U.S.: An Economic Analysis,* p. 239.

[19]After taxes rates of return on equity of U.S. corporations are about 10 percent.

[20]Mead and Sorensen, *op. cit.*

[21]John Menge," The Role of Taxation in Providing for Depletion of Mineral Reserves," *Income Tax Revision,* Ways and Means Committee, Washington, D.C., Government Printing Office, 1951, Vol. 2, p. 970.

[22]Charles Galvin, "The Deduction for Percentage Depletion and Exploration and Development Costs," *Income Tax Revision, op. cit.,* p. 933.

[23]CONSAD Corporation, "The Economic Factors Affecting the Level of Domestic Petroleum Reserves," Part 4 of *Tax Reform Studies and Proposals,* U.S. Treasury Department, Published by the House Ways and Means Committee and the Senate Finance Committee, 1969.

[24]The relationship between changing import prices and U.S. well-head prices is not a one-to-one relationship because of transportation costs. Changing of import prices tends to affect U.S. markets near ports.

[25]*Petroleum Study (Gasoline and Oil Price Increases).* Hearings Before the Committee on Interstate and Foreign Commerce, U.S. House of Representatives, 83rd Congress, First Session, Government Printing office, Washington, D.C., p. 656.

[26]A. Kahn, "The Depletion Allowance in the Context of Cartelization," *American Economics Review,* 54, June 1965, p. 303. See also James Nelson, "Prices, Costs and Conservation in Petroleum," *American Economics Association Proceedings,* 48, May 1958, p. 502.

[27]A. Wright and J. Cox, "The Cost Effectiveness of Federal Tax Subsidies for Petroleum Reserves: Some Empirical Results and Their Implications," in *Studies.*

[28]Some of the leading works that find that capital increases relative to output in general business investments are R. Hall and D. Jorgenson, "Tax Policy and Investment Behavior," *American Economic Review,* 57, June 1967, p. 391; C. Bischoff, "Hypothesis Testing and the Demand for Capital Goods," *"Review of Economics and Statistics,* 51, 1969, p. 354. A critique of this viewpoint is given by R. Eisner and M. Nadiri in "On Investment Behavior and Neo-Classical Theory," *Review of Economics and Statistics,* 50, August 1968, p. 369. For a general review of this literature see G. Brannon, "The Effects of Tax Incentives for Business Investment: A Survey of the Evidence" in *The Economics of Federal Subsidy Programs,* Part 3, Tax Subsidies, Joint Economic Committee, Washington, D.C., Government Printing Office, 1972.

[29]Wright and Cox made use of a new statistical series of prices for natural gas that was not available for the earlier industry behavior studies.

[30]Wright and Cox have considerable doubt about the validity of the CONSAD method of analysis, however.

[31]"The CONSAD Report on the Influence of U.S. Petroleum Taxation on the Level of Reserves," Mid Continent Oil and Gas Association, Washington, D.C., in *Tax Subsidies and Tax Reform,* Hearings Before the Joint Economic Committee, U.S. Congress, July 19,20,21, 1972. Washington, D.C., Government Printing Office, p. 275.

[32]*Oil and Gas Journal,* December 24, 1973.

[33]*U.S. Energy Outlook,* Committee on U.S. Energy Outlook, National Petroleum Council, Washington, D.C., 1972, p. 63.

[34]The time aspect of oil production decisions is emphasized under the heading "user cost" by Stiglitz and by Davidson, Falk, and Lee *op. cit.,* in *Studies.*

[35]Testimony of W.L. Henry for four Petroleum Associations in Hearings on Windfall or Excess Profits, Committee on Ways and Means, U.S. Congress, Wednesday, February 6, 1974, Washington, D.C., Government Printing Office.

[36]"Artificial Restraints on Basic Energy Sources," American Public Power Association, in *Trends in Oil and Gas Exploration,* Hearings before U.S. Senate Committee on Interior and Insular Affairs, August 8, 9, 1972, Washington, D.C., Government Printing Office, p. 1,063–1,073.

Addendum to Chapter Five—The Cox-Wright Model

The tax benefits of percentage depletion and the deduction of intangibles have the effects of increasing the income after tax of oil companies and reducing the cost of adding to the stock of oil wells. The best prospect for reaching conclusions about the effect of the tax benefits lies in investigating past behavior of the oil industry in order to develop a quantitative relationship between some measure of industry size and other variables representing the price of oil and cost of drilling.

Recent work by James Cox and Arthur Wright, building on a number of preceding studies, has reached some impressive results.[1] Their paper develops a quantitative relationship (production function) that explains the level of reserves. In their model this stock of reserves is reduced by production and increased by drilling and equipping new wells. Cox and Wright find that the level of the production prorationing (the state rules limiting production from existing reserves) has a lot to do with the willingness of oil companies to develop new reserves.

From the normal economic assumption that firms should behave in a way that maximizes profit subject to production constraints, Cox and Wright developed an equation that they estimated from industry data from 1959 to 1971. Their equation was

$$\ln R_t = -.34345 + .04856 \ln h_t + 1.15171 \ln Q_t - .16245 \ln S_t - .02994\, t$$

$$(-.15) \quad (4.82) \quad\quad (7.94) \quad\quad (11.28) \quad\quad (6.57)$$

$$R^2 = .9563 \quad\quad DW = 2.1422$$

where,

ln refers to natural logarithms

R_t is aggregate industry reserves of crude oil in the U.S.

h/c represents the cost of capital adjusted for tax benefits

Q is output of crude oil

S is a prorationing factor, and

t is time.

Similar equations were used in which the independent variable was the ratio of the change in R to the level of R and the log of the change in R, with the appropriate change in the statement of the terms for h/c, Q and S. In each form of the equation the coefficient of the output term comes out to about unity, the coefficient of the h/c term comes out to about .05 and the coefficient of the S term comes to about .16. The results of the tests amount to a very strong endorsement of what we have called the output expansion theory of reserves, that is, the tax benefits to the extent that they work do so through slightly lowering the price of oil and gas, which increases sales and output, which in turn requires more reserves. The reserves do not change significantly relative to output but only in proportion to output. The consequence of suddenly finding more reserves is either to reduce other reserve-finding efforts or to increase production.

NOTES TO ADDENDUM TO CHAPTER FIVE

[1]James Cox and Arthur Wright, "The Cost-Effectiveness of Federal Tax Subsidies for Petroleum Reserves: Some Empirical Results and Their Implications," in *Studies.*

Chapter Six

Incentives to Develop Foreign Primary Energy Resources

6.1 THE TAX MAZE

Ostensibly the United States tax law has a deliberate policy of encouraging foreign as well as domestic investment in the production of natural resources. (As a practical matter, the only energy resources affected by foreign investment are oil and gas.) The means of encouragement are the favorable treatment of intangible drilling expenses and percentage depletion, which apply to foreign as well as domestic investment.

For no other provisions in the whole tax system is the comment "things are not as they seem" more fitting.

On a closer examination several circumstances become apparent:

(1) The direct return from foreign investments in oil and gas (except in Canada) is not much affected by the allowance of percentage depletion.

(2) The favorable treatment of intangible drilling expenses is only useful in some highly specialized situations.

(3) The favorable treatment that foreign oil enjoys is connected with some rather technical interpretations of the foreign tax credit rules.

(4) The effect of all foreign tax benefits and foreign tax credit rules sometimes encourages more development of natural resources in countries other than the producing country and sometimes encourages investment in refining and distribution in Europe.

Unlike most of our policy problems, which permit a fairly straightforward comparison of alternatives, analysis in this area has to be reconstructed along the lines of a detective story. Much of the story makes sense only in the light of the peculiar institution of world oil markets, which are dominated by the cartel, the Organization of Petroleum Exporting Countries.[1]

To start with, a little less than one-third of total U.S. foreign investment is in oil and gas, and American oil companies account for about half of the crude oil production in non-communist countries outside of North America. Only a small portion, 11 percent in 1971, of the output of major U.S. oil companies operating abroad was imported into the United States.[2]

Foreign investment in petroleum accounted for about one-third of the book value of total U.S. foreign investments in 1960, but with the faster growth rates of manufacturing, this share has fallen off slightly—to 28 percent. On the basis of commonly cited figures, petroleum investment abroad after foreign taxes has been noticeably more profitable (as a rate of return on book value of investments) than investment in manufacturing.[3]

All of this is a fairly straightforward description. The element of mystery begins when we try to explain the complications of U.S. tax treatment and its interaction with foreign tax treatment.

The central piece in the tax puzzle is the foreign tax credit granted under U.S. tax law. Basically, the credit is a device to avoid international double taxation of the same income. Following the principle that the country in which income is earned has the first right to tax it, the United States permits its citizens and companies earning income abroad to subtract from (credit against) their U.S. income tax the foreign income tax paid on that income.[4] Assume that a firm subject to a 48 percent U.S. tax rate earns 100 abroad and pays 40 of foreign income tax. If this foreign tax were deducted as an ordinary cost of doing business, the U.S. tax would be 48 percent of 100-40 or 28.8 percent. Under the credit approach the U.S. tax is 48 percent of 100 less 40 or 8 percent. Allowing foreign tax to be subtracted from U.S. tax is more advantageous to the taxpayer than if he were to deduct it from income.

There is some controversy about the idea of a foreign tax credit in the first place.[5] We cannot review the whole issue here, but we can observe that the foreign tax credit does imply a generally neutral government attitude toward foreign investment and that to most policymakers this general posture is an acceptable one for a country as rich as the United States. As was explained in Chapter Two an arrangement such as the foreign tax credit is necessary to prevent companies from paying more tax on international business than on domestic business. We accept the general view that there should not be international double taxation. Our concern is with how the foreign tax credit may work differently for oil and gas, that is, that it may go beyond relieving international double income taxation.

Three features of the foreign tax credit rules are particularly significant for oil and gas. The first is the limitation on the amount of the credit. Broadly, the aim of the credit—and the purpose of the limitation—is to allow an offset for foreign taxes on foreign-source income against potential U.S. tax on the same income, but not on U.S. domestic source income. The limitation becomes tricky if there are activities in several countries. Consider the following cases:

	Company I *(Foreign loss case)*			*Company II* *(Low foreign tax case)*		
	Foreign income or loss	*Foreign tax*		*Foreign income or loss*	*Foreign tax*	
		Amount	*Credit*		*Amount*	*Credit*
Country						
A (producer)	100	72		100	72	
B	100	48	0	0	0	
C	(100)	0		100	24	
All foreign countries	100			200	96	
Foreign tax credits						
—per country limitation			96			72
—over-all limitation			48			96

Though the foreign tax credit is limited to 48 percent of foreign income, a taxpayer has a choice of applying the limitation rule country by country or on an overall basis. On a country by country basis Company I gets 96 of foreign tax credit ($48 from countries A and B). On an overall basis Company I has only $100 of net income (due to the loss), so it would get just $48 of credits despite $96 of foreign tax. For Company II the problem is different. On a country by country basis it would be limited to $24 credit in A and $48 in C for a combined 72. By electing the overall basis, Company II can use the fact that the tax in A is low to take credit for the tax in C that goes over 48 percent, that is, a total credit of $96.

For oil companies in less developed countries where crude is produced, tax rates are typically well over 48 percent and, with the percentage depletion and intangible drilling expense deduction, the U.S. tax rate will be well below 48 percent thus, there will be large unused (or excess) foreign tax credits.

If the company is engaged in other foreign businesses, such as refining, transportation and marketing, it might be in the situation of Company II and find it profitable to use the overall limitation—that is, to use some of the unused tax credits against low-taxed income from other countries. For example, the company might organize its oil shipping business in a "tax-haven" country, such as Panama, with low tax rates. Ordinarily a low foreign tax on shipping profits would mean that there would be additional U.S. tax to pay when the shipping profits were brought back to the U.S. The overall limitation permits the unused credit from production to wipe out the extra tax on shipping.[6]

Another common case is one in which the company is also engaged in starting operations in another potential producing area. In this case the intangible drilling expense deduction means that there will be large "losses" in the new country, such as those sustained by Company I in the example. The maximum tax advantage in this case would be to use the per-country limitation on the foreign tax credit and take the new country "loss" as a deduction on the U.S. tax return.

Apart from the intricacies of limitation, there is another feature of the foreign tax credit that has an important application for oil and gas. The theory of the foreign tax credit is most clear when it is thought of as avoiding double application of the similar income taxes of two countries on the same income. The foreign tax credit has never been extended to outright foreign excise taxes because it is presumed that an excise tax is shifted forward to the customer and an income tax is borne by the taxpayer. If a tax has been shifted forward, or added to the price, the obviously correct income tax accounting would be to deduct it from the price, that is, the producer's income, which restores the *status quo ante*. There is no need to make it a credit against income tax.[7] But if a tax falls between the definition of an income and an excise tax, what do we do? (The U.S. tax law extends the credit to a very limited group of taxes in lieu of an income tax, but that provision is not relevant here.)

The tax paid on foreign investments of oil and gas companies in the OPEC countries (where most foreign oil and gas production occurs) is peculiar as an income tax. In the first place, it is not a general tax; it applies only to oil and gas companies. In some of the OPEC countries general business income taxes simply don't exist. Also, the governments in the OPEC countries are effectively the owners of oil in the ground, so that a payment to the host government would look very much the same, whether it were a tax or a royalty.

There is every reason to assert that the bulk of oil company payments to host countries are in fact royalties and that this is relevant to the eligibility of these payments for full foreign tax credit.

Imagine that a foreign country resorted to the following arrangement as a means of attracting investors from other countries to the hotel business: government would (1) pay the wages of bellboys and desk clerks at all hotels, and (2) impose an income tax at double rates on the hotel business. If the extra tax were comparable to the payroll savings, foreign-run hotels would not be put at a disadvantage by this arrangement. The United States Treasury, however, would be put at a disadvantage if this payment to the host country stood up as an income tax. Deductible payroll costs would have been converted into a foreign tax credit.

It is recognized in the public finance literature on foreign tax problems that it is difficult to talk about equality in income taxes in different countries because taxpayers in these countries may be getting different levels of ordinary government services. Our hotel example will serve to illustrate the differences. Say, for instance, that in one country refuse pick-up is provided for foreign-run hotels by the government in return for taxes and that local firms pay separately for refuse pick-up, which becomes a deductible expense. There are no articles in tax journals that we know of arguing that the foreign tax credit should be corrected for different refuse arrangements.

Is the arrangement for paying wages of bellboys and desk clerks different from the garbage collection case? We think the answer must be yes for two reasons. In the refuse case there was also an extra income tax on hotels that makes rather clear the artificial character of the arrangement. Furthermore, a service such as refuse collection is commonly provided free by a foreign government; in comparison, the remarkable character of free bellboy-desk clerk services and the almost exclusive presence of foreign investors in the hotel business would suggest that there was collusion to convert a deductible cost into foreign tax credit.

If the United States wants to preserve its present system of taxing income from foreign investments, it must exercise some prudence in preventing such collusive devices as our hypothetical hotel deal.

On the face of things the tax-royalty situation of the OPEC countries, especially in the Persian Gulf, looks like a collusive device.[8] The general economic principles of rent-royalty income, which are well established, tell us that rents and royalties on different lands should be related to differential production costs.[9] Production costs in the Persian Gulf are reliably estimated as under 10 percent of costs in the United States,[10] but the nominal royalty rate (12.5 percent) is even less than royalty the rate in the United States.

A further theoretical problem with the application of the foreign tax credit to the taxes of the OPEC countries is that they are not strictly income taxes even in form (putting aside the royalty issue). The taxes are levied on hypothetical gross income, calculated from a posted price, reduced by a hypothetical cost. It is not uncommon for royalites to be related to gross income.

The peculiar structure of OPEC income taxes arose from the concern of OPEC governments about sales of oil at discounts from posted prices. When an OPEC country is dealing with a producer country that has refining and distribution activities outside of the OPEC country, it will be to the advantage of the producer to sell to its refinery and distribution affiliates at artificially low prices in order to reduce the amount of profit exposed to the high OPEC tax rate.[11] Currently, sales seem to be at about 80 percent of posted price in the Persian Gulf.

The royalty argument and the excise tax argument for disallowing the foreign tax credit for OPEC taxes are closely related. As we argued earlier, the reason for noncrediting of excise taxes is that these are presumed to be shifted forward to the price. The economist's argument that a corporate tax is not shifted forward in price, applies only to a general income tax.[12] Oil, however, is an international commodity, and the high OPEC tax is clearly something to be added to the price. The charge announced in Saudi Arabia in late 1973, for example, was twice as high as the previous market price! Clearly it was assumed that the producer companies would raise prices so that company income per barrel would not be reduced.

On the grounds of tax theory, therefore, the extension of the foreign tax credit to the OPEC charge is highly questionable.

6.2   THE EFFECTS OF THE TAX MAZE

The foreign tax provisions of U.S. tax law as they have been applied to oil and gas turn out to have unexpected results. Much of this outcome occurs because the foreign "tax" is so high that it would exceed the United States tax on income from foreign oil and gas production even if the firms were not allowed percentage depletion or deduction of intangible drilling expenses in computing their U.S. tax. This was partly true prior to 1973 when the posted price was about $2.30 per barrel, the "tax" $1.12, the market price about $1.86 and the company margin about 30 cents.[13] In 1974 when the posted price in Saudi Arabia went to $11.65, the "tax" to $5.54 and the company margin to the vicinity of 50 cents, the results were even more striking.

The specific effects of this structure are hard to pin down because we do not have precise data on the outlays for foreign drilling by U.S. oil companies; however, three distinct patterns can be identified. The first applies in nearly all cases. The other two are alternatives, and we can only say that some companies are involved in Pattern 2 and some companies are involved in Pattern 3.

Pattern 1—There is no net U.S. tax on the net margin that

the companies earn from production of oil in the OPEC countries be-cause the companies are allowed a foreign tax credit for what the OPEC countries call an income tax.

Pattern 2—Some oil companies, apparently those with only limited drilling programs, can use the excess foreign tax credit from production to credit against U.S. tax on income arising outside of OPEC countries from refining, transportation, and marketing. These companies then pay a total foreign tax at less than the U.S. rate. This is the pattern of Company II in our previous illustration, and it comes about particularly because of the low foreign taxes on shipping subsidiaries.

For a company in Pattern 2, we can estimate that each barrel of oil generates a 50-cent profit on production and that there is an additional 50-cent profit on the other processes, which incur about 13 cents in foreign income tax. The other income of 50 cents a barrel would incur a U.S. tax of 24 cents. Were it not for the excess foreign tax credit on the production operations, there would be a net U.S. tax of 11 cents a barrel on the other income. By avoiding additional United States tax on these other activities, the companies serve to divert capital funds from the United States on terms that amount to a subsidy to European and Japanese oil consumers.[14]

Pattern 3—Some oil companies that have extensive drilling programs may do a great deal of drilling in a country in which they do not have much established production. In these "new" countries there are likely to be extensive start-up losses; the companies therefore choose to forfeit the opportunity to use the overall limitation on the foreign tax credit in order to take deductions against their U.S. income tax base for these losses. When the new country operations are established, it is likely (unless the foreign country has a net loss carryover provision) that the foreign tax credit provisions will come into play; thus, the com-panies avoid tax on the profits then being generated. The lopsided result is that the U.S. Treasury pays for the early losses but gets no benefit from the later gains. This is a clear subsidy for operations in potential new pro-ducer countries.

There are several kinds of changes in the U.S. law affecting taxes on income from foreign oil and gas operations that should be considered in relation to these three patterns. In the first place, the tax in oil-producing countries is so high that if the foreign tax credit provisions continue to work as they do now, it will make no difference for foreign oil operations whether percentage depletion is continued or not, or whether it is disallowed for foreign operations.[15]

If the United States repealed the deduction for intangible drilling expenses this would not affect income from producing countries

(Patterns 1 and 2), but it would reduce the size of the start-up loss advantage involved in Pattern 3. Another way to deal with Pattern 3 would be to provide that income from companies with oil and gas operations in a new country would not be eligible for any foreign tax credit until the company had paid U.S. income tax on an amount of income as large as the prior aggregate losses.

To deal with Pattern 2—the use of excess foreign tax credits to reduce additional U.S. tax on such foreign activities as refining transportation and marketing—we could either repeal the general rule, the so-called overall limitation that permits the transfer of foreign tax credits to operations in other countries, or we could deny the application of this rule to excess credits arising from oil production.

There is considerable literature arguing for elimination of the overall limitation on the foreign tax credit in any situation. Such a change would prevent the excess tax in one foreign country from being used to reduce the U.S. tax liability on a distinct operation. We think that if a foreign country wants to over-tax a U.S. company, then it should be up to the company to decide whether the benefits it receives are enough to justify continuing to operate in that country. The current U.S. tax policy only finances and encourages over-taxation.

Without rejecting the overall limitation in all cases, one could decide that any tax credit allowed for the "income" tax, whether considered a royalty or excise tax, in the OPEC countries would not be allowed as credit against income from other countries. This is the current recommendation of the Treasury Department.[16]

Treasury's technique would be to deny the percentage depletion allowance on foreign oil and at the same time allow oil companies to treat as an income tax only as much of the OPEC payment as exactly wipes out tax on production income. To illustrate: If the market price (less cost) were $7.50 per barrel and a host country takes $7, Treasury would deem $6,54 a barrel as royalty income and 46 cents as tax. Deducting the royalty leaves a before-tax income of 96 cents on which the U.S. tax would be exactly 46 cents, leaving no excess foreign tax credit.[17]

From this it can be seen that dealing with Pattern 1—U.S. tax on production income—could involve either denying the foreign tax credit for the OPEC charge completely or deeming more than $6.54 of the total payment as a royalty and less than 46 cents as tax. The extreme of disallowing the foreign tax credit completely would mean applying a 24-cent net United States tax to the 50-cent margin. Depending on the specific royalty rule adopted, the net United States tax could be made to come out anywhere between 0 and 24 cents a barrel.[18]

Of fundamental importance is a question concerning Pattern

1, "How would things change if the United States started imposing a net income tax on the margin from production in the OPEC countries?" We think that the initial, short-run effect would be an increase in the price of oil in the world market of approximately the amount of the tax. The market situation seems to be that short-run demand is very inelastic, which accounts for the relatively free manner in which OPEC has engineered price increases.

There is no assurance that there would be a price increase to cover the full cost of the additional tax on U.S. producer companies. At present the OPEC countries do not directly control the market price. Their direct control is limited to the take. The host countries are not, however, indifferent to the price.

In January 1974 Iranian Finance Minister Jamshid Amouzegar was quoted in connection with reports of occasional sales of crude at very high prices:

> We get blamed for all the fuel price increases, but the companies that buy out oil for $7 a barrel are selling it at $15 a barrel. We believe that the oil companies should get only 50 cents a barrel.[19]

Further complicating the situation is that parent countries, other than the U.S., of companies producing oil in the OPEC countries do not impose significant taxes on the margin.[20]

In the short run the world market price of oil should rise simply because the non-United States companies would find it difficult to increase their output; consequently, there would be little short-run loss of markets by U.S. firms, even though they would be passing on the tax increase. If the other parent countries, such as the United Kingdom, France, and the Netherlands, did not impose higher taxes on their companies, these companies would have a long-run advantage because they could enjoy higher prices without higher taxes.

If there were a full price increase to offset the increased United States tax, it would be twice the amount of the tax itself, because the increase provides more taxable income. One short-run effect of our generosity in the present foreign tax arrangements is to maintain a lower oil price for Europe and Japan, the principle consumers of Middle East oil. There is no advantage to the United States in subsidizing foreign oil prices.

Whatever the price effect of an increased tax on production income by the United States, if the tax were not duplicated by the other parent countries, the situation would be unfair to the U.S.-owned companies, which might suffer losses by being thrown into an unfavorable competitive position. In view of Jenkin's conclusions that returns on overseas oil and gas investment are' somewhat above average for foreign investment, however, the fairness consideration does not preclude some modest changes in the U.S. tax law now and larger changes introduced gradually.

Assume that by a series of gradual changes, the U.S. tax on production income was increased and that the tax generated a significant competitive advantage for other foreign-owned companies so that the portion of world oil produced by U.S. companies fell sharply. This would be a bad situation for the owners of U.S. companies, but would it be a disadvantage to the United States?

This question has, over the years, been at the root of much of the debate about how the tax law should treat foreign income. Businessmen generally argue that the tax law should strive for parity between the U.S. foreign operator and the other foreign companies with which it must compete. Economists have generally argued that the parity should be between the U.S. foreign operator and the U.S. company that operates at home. The economists' argument is based partly on the assumption that the United States is better off if investment stays at home.

Additional investment generates three conspicuous income flows: more profits to the investor, more profits taxes to the government, and higher wages for workers due to increased productivity. When United States capital is invested abroad, only the additional investor profits accrue to Americans. The profits taxes and the increased wages accrue to foreigners.[21]

We can illustrate the general argument that the United States should not worry about the competitive position of U.S. companies—beyond the problem of fairness caused by sudden changes in the rules—by giving an example. Assume that U.S.-owned companies are engaged in producing computers in France and that France decides to subsidize the computer business by imposing on it only half the regular tax rate. If the United States does not have such a subsidy in its own law—because it sees no advantage in distorting market decisions about the relative share of investment that should go into computers—then it should not make capital available to provide over-investment in computers in France.

When we transfer this argument to the oil business, the question becomes "Is the United States better off if oil in, say, the Persian Gulf or in Venequela is pumped by American companies"? We would expect that whatever the nationality of the company, it would import oil into the United States as long as price relationships made it profitable; this must be why, for instance, British Petroleum sells oil in the United States now. On the basis of scanty press reports about the "leakages" around the Arab oil embargo in the winter of 1973 and 1974, it does not appear that U.S. companies are responsible. Less stringent enforcement in particular host countries seems to be responsible. We have very little solid information on all this, however. Possibly it does help the United States to own a lot of the pumping companies. The evidence is not clear.

6.3 POLICY RECOMMENDATIONS

We began this chapter by describing the foreign tax provisions related to oil and gas as a maze. The problems growing out of this maze are no simpler, and there are no easy solutions. One way of approaching solutions is to organize our recommendations around three basic questions.

(1) *Should the crediting of OPEC "income" taxes permit U.S. oil companies to reduce United States taxes on other foreign business that they conduct?* Our answer to this is no. (It is the Treasury's answer also.) This could be done by removing the overall limitation on the foreign tax credit; thus, all companies would have to look at the creditability of the foreign tax on an individual country basis. (The Treasury alternative of a royalty formula designed to prevent carryovers is generally acceptable in the context of the present work, which is only concerned with taxation of natural resource industries.) The reason for our position is, briefly, that the benefits of this low taxation flow mostly to foreign customers.

Critics of our agreement will say that some of this benefit goes to United States shipping and that we have historically followed a policy of subsidizing shipping. Our response is that if shipping subsidies are needed, then they should be specifically justified in the budget as an expenditure item.

(2) *Should U.S. oil companies operating abroad continue to enjoy the present advantageous treatment of start-up losses?* The present rule under which the companies deduct large "losses," which are really investments and never restore the offsetting income, should certainly be ended. The question is how far to cut back. We would favor a rule such as the one proposed by Treasury: When start-up losses have been deducted for U.S. tax purposes, then foreign tax credit should be denied until an amount of income has been included on U.S. tax returns equivalent to the prior deduction. (Until this offset has occurred, there has been no net foreign income from that country.)

Our earlier recommendation to terminate the deduction for intangible drilling expenses would also sharply reduce the amount of start-up losses. Even if we do not eliminate the intangible deduction in the United States, should we eliminate it for foreign drilling? Several tax-law incentives are now limited to domestic investment.

We think that the discovery of new oil deposits in a country that is not now in OPEC, or the discovery of new reserves in an OPEC country that is only a small producer, could be very advantageous for the United States. Long-run maintenance of an OPEC price requires limitation on production. To prevent market disruption in the face of new discoveries, existing OPEC producers would have to reduce outputs or forgo some growth in output to accommodate the newcomer. These accommodations would be uncomfortable, and the outcome could well be that the newcomer would be better off selling oil at slightly under the cartel price with no output limitation.

We would like to break the monopolistic hold of the OPEC cartel, but it would be best to do it by specific subsidy arrangement, not with the intangible drilling expense benefit. The subsidy could be negotiated so that there would be some opportunity to impose conditions on the host country. Moreover, we would get more for our money under a subsidy.

The intangible drilling expense deduction is, as the American Petroleum Institute told the Ways and Means Committee, "success oriented."[22] This emotionally loaded term was clearly chosen to elect a sympathetic response from the Committee, but fundamentally it is an argument against the provision. Success orientation means that the less incentives are needed, that is, the more success is assured, the greater the tax incentive will be. Deductions of intangibles for successful wells is a very large subsidy when a company has discovered oil and is developing a field preparatory to going into production. At this stage no incentive is needed. If incentives are needed, they should be given at the exploratory stage, possibly in the form of low-interest loans.

(3) *Should the United States impose a net tax on the production margin in the OPEC countries?* This is the hardest question of the three. There is a great deal of money involved. At a margin of 50 cents a barrel and with output by U.S. companies of 6.5 billion barrels, we are talking about over one billion dollars of U.S. tax. If, as we expect, the margin would rise to offset the tax, then there would be even more margin to tax, and the revenue could approach 2.5 billion dollars.

There are a few reasons for not imposing tax on the production margin that we should consider before reaching a conclusion:

a. If we base our tax on the production margin on a tax legality that the OPEC charge is not eligible for the foreign tax credit, there is no bar to the OPEC countries revising their tax laws so that they would meet U.S. tax law requirements. The OPEC countries could, for example, continue to base their royalty charge on an artificial posted price and raise the royalty rate to about 55 percent. Then a tax that looks like a real income tax could be imposed at a rate of 48 percent on the receipts minus royalty and expenses. By using the posted price in the royalty formula, the countries would protect themselves against unfavorable transfer prices. Furthermore, eliminating any carryover of unused foreign tax credit from OPEC operations (our answer to Question 1) would mean that there would be no advantage to the companies in designating extra payments as tax, which was the case in the 1950's. OPEC could with little cost design a tax that we would have great difficulty disallowing under present law.

b. If we did succeed in imposing a tax on the production margin of U.S. companies abroad, this would, as we have seen, work to the competitive disadvantage of U.S. companies whether or not there was a price increase. It is also possible that a smaller share of Middle East production under control of American companies would be some disadvantage to the United States.

In view of comment "a" above, it seems to us that it is fruitless to expect to settle the present problem on the basis of legal theories of a tax credit. We need to go directly to the question of results. There is a virtual economic war going on between the OPEC and the consumer nations. Would enacting a different foreign tax credit law, for example, one that specifically denied a credit for OPEC taxes, help the United States? We think that it would, if all the home countries of the producer companies took the same action.

From a long-range standpoint, the price strategy of the OPEC cartel must be to maintain an oil price somewhat below the level at which it will be profitable for consumer nations to turn to substitutes. This is an oversimplification of the OPEC price strategy, but it is adequate to establish the point that there is a ceiling and that the maximum OPEC take is the ceiling less the cost of production, including a normal profit for the producer companies. If the home countries of the producer companies all imposed a tax on the producer companies, it would effectively increase their cost of operation, and with the long-run ceiling on the OPEC price, the increased tax must come out of the OPEC take. We think that the United States should initiate discussions with the other home countries on common action to impose home-country taxes on the production margin because of the very special problems of the oil cartel. Agreement on this would not be easy. In the short run it could be expected that the tax would increase the price. European countries and Japan would suffer financially because their share of international oil is larger than their share of ownership of production. Their payments would increase more than their tax receipts. In the long run, however, the European countries and Japan would gain, based on our analysis that the burden of the tax would come out of the OPEC take.

Another area of uncertainty raised by the proposal for unified tax action involves the possible OPEC responses. There is much talk of nationalization of foreign oil companies, after which the present producer companies would be employed on service contracts. There is also much speculation that the OPEC cartel would find it harder, in these circumstances, to maintain price discipline.[23]

The present paper cannot take into consideration a full analysis of the political ramifications of the preceding tax proposals, but it does appear to us that using taxes is one way that the consumer countries, which are substantially the same as the investor countries, could fashion a counter-weapon to the power of the cartel. It seems worth the effort to explore this at the diplomatic level.

NOTES TO CHAPTER SIX

[1]The comprehensive study of the international situation in oil is by M.A. Adelman, *The World Petroleum Market, op. cit.*

[2]The source of figures not explicitly cited is Glenn Jenkins in *Studies.*

[3]Calculations of rates of return on capital are notoriously unreliable due mostly to the fact that "book" values of corporate assets do not reflect current market or replacement values. Jenkins has done some careful adjusting of the published income and figures to remove bias and concludes that petroleum investment has indeed been more profitable. Glenn Jenkins, *ibid.*

[4]Many foreign countries simply refrain from taxing foreign income, but one way or another there is a general effort to avoid double taxation and let the host country tax first—at least as far as business and natural resource operations are concerned.

[5]For an academic critique, cf. Peggy Musgrave, "Tax Preferences to Foreign Investment," *The Economics of Federal Subsidy Programs,* Part 2, Joint Economic Committee, U.S. Congress, 1972. At a more popular level, there has been extensive literature favoring the Burke-Hartke bill, which would deny the credit.

[6]In 1969 Congress modified the foreign tax credit rules to prevent any excess foreign tax credit from arising because of percentage depletion being used to provide foreign tax credit for income not related to the oil business. Due to the increase in the OPEC tax in 1973, this provision has no effect. There are ample unused credits anyway.

[7]Cf. Peggy Musgrave, *U.S. Taxation of Foreign Investment Income,* Cambridge, Mass., Harvard University Press, 1967, pp. 111–115.

[8]For an assertion that the U.S. government participated in denominating the OPEC charge in a way to qualify for more foreign tax credit see J.E. Hartshorn, *Politics and World Oil Economics,* New York, Praeger, 1967, pp. 198–200.

[9]See Chapter IV above.

[10]M.A. Adelman, *op. cit.,* p. 76.

[11]*ibid.*

[12]For a discussion of the difference in "shifting" between general and special income taxes, see Otto von Mehring, *The Shifting and Incidence of Taxes,* Homewood, Ill., Richard Irwin, 1942.

[13]Price data is for Arabian light 40, 1.7 percent sulfur. It is taken from *Petroleum Intelligence Weekly* (various dates). The company margin estimate is by the Department of Commerce.

[14]If half of the U.S. company share of foreign production is involved in the pattern, the aggregate amount of the subsidy could be in the neighborhood of $350,000,000.

[15]Canadian operations of U.S. owned companies are an exception because the Canadian law provides a benefit for oil production similar to our percentage depletion. If percentage depletion were repealed in the U.S., we think that Canada would follow suit.

[16]See testimony of Secretary Shultz before the Ways and Means Committee, February 4, 1974.

[17]These numbers refer specifically to the Saudi Arabian take on Arabian light oil. The amounts would be different for other types of oil and for other countries.

[18]This range somewhat overstates the upper limit of U.S. tax because we have made no specific deduction for drilling. The data is not available.

[19]*Washington Past*, January 9, 1974.

[20]Based on tax law summaries provided by the American Petroleum Institute.

[21]Cf., P. Musgrave, "U.S. Taxation of Foreign Incomes" *op. cit.*

[22]Testimony of American Petroleum Institute before the Committee on Ways and Means, February 6, 1974.

[23]M.A. Adelman, *op. cit.,* pp.

Chapter Seven

Taxes and Subsidies and the Public Utilities in the Energy Field

7.1 THE RANGE OF PROBLEMS

By public utilities in the energy field we mean principally companies concerned with the generation and distribution of electricity or with the distribution of natural gas.[1] Practically speaking, these firms are not concerned with the availability of resources but with marketing energy. It will be helpful to think of the generation of electricity as simply a technique for marketing the energy content of coal, oil, and uranium. (The hydrogeneration of electricity is a very small element of the total energy picture.)

If we accept the fact that problems of availability of natural resources have been dealt with by other policies, then policy problems regarding public utilities and the energy crisis are of a different order of magnitude than, say, the prospect of running out of natural gas. The problems that may creep into the marketing sector of the energy process relate to inefficiencies.

We are not dealing with all of the public policy issues related to public utilities in the energy field, but only with those that are potentially related to tax and subsidy questions. There is some value, however, in explaining the relationship between our particular analysis and the toal problem.

In the public utility area, the most conspicuous manifestation of the energy problem is interruption or curtailment of electric service, which has happened in the past. By and large, this is a planning problem unrelated to tax-subsidy policy. The government has involved itself in investment planning in the utility industry in a number of ways, including federal price controls, rate regulation by public utility commissions, and environmental regulation affecting siting and construction.

Government has also affected private investment in unintended ways. As the federal government urged ambitious development plans for nuclear energy, it caused private companies to see less room for new investments in conventional electricity generating capacity. It has turned out, however, that the ability to generate electricity from nuclear plants has been appreciably less than the government has led private utilities to expect. This planning error has left private utilities with somewhat less than the total capacity they expected to have.

The same kind of problem arises from environmental constraints imposed on plant selection and design. Planned expansion of plant capacity has frequently encountered opposition from various environmental groups; the process of considering the environmental objections and finally approving the capacity expansion or developing alternatives has involved considerable delay.

As long as the federal government pursues an active role in planning electric capacity expansion, there will be potential for planning errors and shortages. Although this is not a tax or subsidy matter, we can offer the general comment that the consequences of planning uncertainties can be minimized by following fairly regular planning procedures and undertaking changes slowly. With regard to new technologies, such as nuclear generation, it should be expected that outcomes will differ from plans. These differences can, however, be dealt with out of normal reserve capacity if we do not plan to meet large portions of the capacity demand with unknown techniques. As we learn more about techniques, we can make changes faster.

A second broad area of public utility problems is directly related to tax-subsidy questions. The transformation of natural resources into electrical energy causes pollution, which is conspicuous in the generation of electricity from uranium and from high sulfur fuel. The pollution control as now practiced involves government regulations that increase the cost of generating electricity. Questions can be raised about the techniques of enforcing environmental requirements, especially alternative uses of regulations and effluent taxes, which are dealt with extensively in Chapter Eight. These issues also involve important income-distribution effects, dealt with in Chapter Ten, depending on whether the cost is met by public subsidies or passed on to consumers.

Finally there are problems associated with the economic structure of the public utility industry, which are dealt with in part in another study in the Energy Policy Project series. Some of the following questions are involved; they cannot be readily separated from a discussion of taxes and subsidies:

1. Whether the system of price regulation leads to excessive use of electricity and consequently to an inefficient drain of the limited fuel resources;

2. Whether excess use of electricity is related to the presence of a considerable amount of public ownership of electricity generating systems;

3. Whether the structure of price regulation leads to inefficient use of resources.

Probably the major functional difference between public and private ownership is the tax advantage associated with public ownership. A major issue in the existing price regulation of privately owned utilities is connected with the potential of regulation to lead to an excessive use of capital in electricity generation and distribution. The situation is aggravated by the fact that the federal tax is taken into account in the price-setting process.

The federal tax law involves some special incentives for additional capital investment, such as the investment credit and accelerated depreciation rules, both of which operate somewhat differently for public utilities than they do for other firms. One must ask whether it is sensible from an energy-policy standpoint to provide different treatment of public utilities in the investment tax incentives. This question, in turn, can only be answered in the light of an analysis of the complex ways in which the existing price regulation bears upon investment in public utilities.

7.2 FEDERAL INVESTMENT TAX INCENTIVE POLICIES

The federal government has taken a position on several investment tax incentive questions with regard to privately owned public utility companies. The first related explicitly to the investment incentives extended to business generally and how they are applicable to public utilities. In this regard, Congress has allowed public utilities to make use of the same accelerated depreciation that other business taxpayers use, but it has allowed only three-sevenths of the investment credit given to other taxpayers.

The other policy decision about investment incentives concerns how utilities will be treated for rate-regulation purposes. In principle, the utilities are entitled to a fair rate of return after income tax. If the investment incentive, such as investment credit, is regarded as a reduction in the federal tax, as a tax rate cut is, it should be passed on to customers in the form of a price reduction. The price reduction, with a 48 percent tax rate, means still-lower profits, more tax reduction, and more price reduction. In aggregate, the price reduction is about two times the amount of the investment credit. The same would be true of a tax reduction attributable to accelerated depreciation. For many businesses the incentive aspect of accelerated depreciation is complicated by the fact that, even though federal income tax allowances for depreciation are generous, there is no clearly defined figure for depreciation and no figure for the incentive. Public utilities, however, show on their books and in public reports a separate depreciation cost, which is used for rate-making, even when they use accelerated depreciation for tax purposes. The investment incentive, or the tax saving from using artificial tax depreciation, is quite explicit for public utilities.

Another way to look at investment tax incentives is to consider that the company paid its regular tax and that it received a federal subsidy equal to a percentage of the purchase price of machinery and equipment. The thrust of this view is that none of the tax reduction due to the investment incentive should be passed on to customers but instead should be available to the utility in a way that brings about a change in its investment policy, which investment incentives are supposed to do.

The accounting-regulatory technique that involves using investment tax incentives to reduce price is called "flow-through." The alternative to flow-through is called "normalization." In practice, when state utility commissions have permitted normalization, they have recognized that when the tax incentive is used to pay for more capital equipment this much of the capital should be regarded as investment of public monies and the company should not earn any rate of return on the tax savings from investment incentives.

State regulatory commissions have followed different policies with respect to normalization and flow-through. Gradually, in the 1960's, the Federal Power Commission moved in the direction of requiring flow-through for both the investment credit and the benefits of accelerated depreciation. In both cases, Congress has responded by amending the tax law in an awkward way, but one that indicates a Congressional intention to favor normalization.

What is ultimately at stake in both the FPC and the Congressional decisions is the matter of how the investment treatment should differ between regulated utilities and normal businesses. In 1962 Secretary of the Treasury Douglas Dillon testified before Congress that investment incentives were not required for utilities because they had to make investments to meet their service obligations and the rate of return on these investments were fixed by regulation.[2]

The view to the contrary was well articulated by Mr. Donald Cook, President of the American Electric Power Company. Cook argued that regulated utilities did apply a profit and cost calculus to marginal investment decisions and that an investment incentive was as likely to change decisions at the margin in the electric industry as elsewhere.[3]

Considering the broad level at which the issue was debated in 1961 and 1962, the best industry could salvage was a committee compromise to provide three-sevenths of the investment credit for utilities.

On careful analysis, there is a more subtle point at issue than that involved in either the industry or Treasury argument. Treasury assumed that public utility commissions tell the companies how much to invest. The industry argument assumed that public utilities make their own investment decisions on a profit calculation, just as any other private business does.

As students of public utility economics have come to recognize, neither of these views is adequate. The commissions cannot dictate the amount of investment but they can generally control the rate of return. In view of the commissions' power, privately owned public utilities cannot operate as other businesses do. They must accept what the commissions fix as data or constraints and operate as profitably as possible subject to the constraints.

The outcome of analyses of public utility operations as profit-seeking subject to constraints has been a body of articles on the "Averch-Johnson argument."[4] In its simplest form the Averch-Johnson thesis is based on the theory that in a public utility there can be various technological combinations of capital, fuel and labor to produce any given amount of electricity. Capital, then, can be subsituted for labor by designing more intricate, automated machines or capital can be substituted for fuel by designing machines with greater heat efficiency— by incorporating recirculation of steam to prevent heat loss, for instance.

With a range of possible levels of capital investment for a given level of electrical output, one would expect the economic law of diminishing returns to apply, that is, the marginal increases of efficiency from using more and more capital would decline. Assume, for example, that a given amount of electricity output could be produced with 10, 11, or 12 units of capital. Assume also that the return on capital earned with the regulated price and 10 units of capital is 9 percent, that the marginal return from adding the eleventh unit of capital is 7 percent and that the marginal return from the twelfth unit 5 percent. Finally, assume that the regulatory commission establishes 7 percent as the fair rate of return on utility capital. To simplify the example, we also treat 7 percent as the real cost of capital.

From the facts given above, it would seem that ideally the electric company should invest 11 units of capital because at this point the marginal return on the eleventh unit is equal to the allowed return, 7 percent. (In reality, the capital intensity implicit in using 12 units of capital for this electricity output is inefficient because the twelfth unit only provides a 5 percent return, but we have specified that the real cost of capital is 7 percent.)

The Averch-Johnson thesis holds that an electric company will tend to operate with 12 units of capital rather than 11 because the effect of rate regulation is to deny the company the return it could potentially make on the first 10 units of capital. Because the electric company should try to operate as profitably as possible under the constraint of earning an overall 7 percent return, this strategy will lead to operating with more than the most efficient level of capital. The company would, in effect, use the excess return on the first 10 units to cover the deficient return on the twelfth unit and end up making the allowed fair rate of return on 12 units rather than on 11, which makes the company better off. The essence of the Averch-Johnson thesis is that industry profit-seeking under the regulatory constraint will lead to over-investment in capital. This does not mean flagrant waste or gold-plating but rather a general tendency to go too far in substituting machinery for labor and fuel.

Students of public utility regulation have not been able to produce much data to prove the extent of this over-investment in capital, though some recent work suggests that over-investment could be 25 percent or more.[5] If these estimates are to be taken seriously, the implication is that it would be wise to deny completely the investment credit and accelerated depreciation for public utilities. These two tax incentives add up to not over a 15 to 20 percent investment incentive. The handicap of not having this incentive would tend to lead to less capital investment in public utilities, but not so much less that the full Averch-Johnson effect would be offset.

More research is needed on the matter. Despite the statistical work that has been done on the problem, testing the Averch-Johnson effect is complex, and there may be some substance to the contentions of public utility executives that they pursue efficiency for its own sake. The present state of our knowledge would suggest caution, although the statutory compromise of a half investment credit appears to be too cautious. Because public utilities qualify for the full accelerated depreciation, a complete disallowance of the investment credit would be appropriate.

Another tax issue that applies to privately owned public utilities is the provision in the tax law that pushes the regulatory commissions in the direction of allowing the tax benefits of the investment incentives to be normalized, that is, retained by the firm rather than passed through to customers.[6]

The issues here are more complex than one would guess by the descriptions "retained by the firm" and "passed through to customers." First, from the customer's standpoint, having the tax benefits of investment incentives retained by the firm means that they will serve to pay for part of the plant capacity. The company will not be permitted to earn a profit on this capacity, and it will bear no income tax. In the long run, having investment incentives retained by the firm produces lower electicity prices. But customer spokesmen in public utility rate hearings say they want lower rates *now*, not in the future. Their impatience is understandable, but customers on the whole would not suffer much if the price reduction comes later and is larger.

From the company standpoint, it looks as if the company is better off keeping the tax benefits of the investment incentives. It was the assumption of the Congressional action that only in this way would the intent of the statute, to encourage investment, be achieved. This is not necessarily the case. A full analysis of alternate consequences, in the light of the Averch-Johnson approach of investigating profit-seeking under a regulatory constraint, suggests that the companies would be better off in the long run by using flow-through.[7] Flow-through for lower prices and more sales and, in turn, more capacity leads to higher long-term profits.

The issue of whether consumers or companies benefit more from flow-through or normalization turns out to be a very close question. On balance, it would be wise to get away from congressional tax committee involvement in public utility regulation. The net balance of benefits involves judgments about the public aspects of the situation that should not be made in tax law.

Before leaving this topic, we should sum up our conclusions and take account of their relation to energy policy. We have recommended that the investment credit now extended on a three-sevenths basis to electricity and gas distribution facilities be eliminated altogether for these utilities.[8] We also recommend the elimination of the tax law provision that attempts to require public utility regulatory commissions to allow normalization of investment incentives. On balance, the two changes would probably have little effect on investment.

The recommendation for eliminating the investment credit moves slightly in the opposite direction of the popular notion of conserving fuel. The thrust of the Averch-Johnson thesis is that, under present arrangements, utility companies use production techniques that save fuel but waste capital. Our recommendation to reduce the waste of capital would imply using more fuel.

No matter how enthusiastic someone is about fuel conservation, he should not be in favor of spending $10,000 to insulate a house when insulation would reduce annual fuel costs by $10. Although this is obviously an extreme case, it should make the point that, though conservation is important, we still want to retain efficiency. Our argument in this chapter has proceeded on the basis that relative efficiency can be inferred from market costs along with profit-seeking behavior. From our general discussion in Chapter One, it should be recognized that relative efficiency would follow if there were no price defects in the market.

The strongest claim for emphasis on fuel conservation is that the market prices of fuel are already too low—because of tax subsidies and current price controls, for example—and that fuel users should not follow market efficiency tests but should prefer wasting capital to wasting fuel. The approach of this whole report, however, is that our lives will become very complicated if we try to get to the right result by pursuing two wrongs. If we persist in having fuel prices that are too low, how can we possibly implement a policy of wasting capital to just the right extent to offset the waste of fuel that is involved in the low fuel price? The only program that seems to have any chance of success is to try to get both of the prices right in the first place.

7.3 PUBLICLY OWNED UTILITIES

The present tax law provides very substantial advantages for publicly owned electrical power companies.[9] These advantages arise from the absence of a tax on the income of publicly owned utilities and from the availability of tax-exempt municipal bond financing for these utilities. These tax advantages are supplemented in the particular case of rural electrific cooperatives by direct federal loans at very low interest rates.

The movement toward publicly owned power was considerably accelerated in the 1930's with large federal investments in projects such as TVA. The literature surrounding TVA articulated quite specifically that the objective of this investment was to provide a "yardstick" to prove to the rest of the electrical industry that consumer demand was quite elastic and that a low price strategy would pay off through higher sales.

Conspicuously, the relative role of public power versus private power has reached something of an equilibrium, at least since the 1950's.[11] This equilibrium, which has been observed in a fairly steady low level of public ownership in the electric business, is a very curious result if in fact publicly owned utilities operate as efficiently as privately owned utilities.

A direct comparison of efficiencies is difficult because most publicly owned plants are small local distributors or are limited to generation of electricity, as is the case with TVA. Direct average cost comparisons suggest that the costs of public power are higher, but this may be due to technological differences between the two classes of firms. The best judgment appears to be that, on the average, publicly owned utilities are in fact operated somewhat less efficiently than private utilities, otherwise we would observe a stronger trend toward public utilities in view of their significant tax advantages.[12]

On the face of things, the allowance of federal tax advantages for publicly owned electric companies is an unwise energy policy. It is a straight-out subsidy that should, assuming equally efficient plants, generate real income advantages to people who live in privately owned power areas. In addition, the passing on of tax advantages in the form of lower prices, which seems to be a typical pattern, means that the consumption of electricity—using up fuel—and the generation of electricity-related pollution are all increased.

It would be very hard to change the tax law to impose a direct income tax on public authorities running electric companies. This could stir up constitutional sensibilities. A further difficulty of imposing a direct income tax on public authorities is that the tax could be avoided by reducing prices and running the authority closer to a break-even point with new capital supplied from local tax funds or bond funds. A common instance of a low-price strategy is the practice of publicly owned power companies selling, say, power for street lights to the parent government at less than cost.[13] Ultimately, the low-price strategy would even further increase the demand for electricity in public power areas. It would thus be poor energy policy.

The technical problem of imposing an income tax on the operation of publicly owned utilities could be avoided by making publicly owned utilities subject to an excise tax that would produce revenue equivalent to what would be produced by normal pricing. There is constitutional precedent for making a state government subject to a general excise tax, which applies to other firms selling comparable products, but at issue here is an excise tax that would be a substitute for an income tax.

Justice Marshall's stirring dictum that "the power to tax is the power to destroy" was originally delivered in a case involving intergovernmental tax immunity.[14] In the O'Keefe case in 1936, however, the Supreme Court removed the barrier to one government imposing income tax on wages paid by the other on the grounds that a tax that is applied uniformly does not put a government at a disadvantage as a competitive employer.[15]

The more relaxed view of intergovernmental immunity reflected in decisions following the O'Keefe case would appear to justify the prediction that the Court would uphold an excise tax, equivalent to the general income tax, on publicly owned utility companies.

The Supreme Court has upheld the application of an excise tax by the federal government on a service provided by another government where the tax did not discriminate.[16]

An excise tax is, nevertheless, different from an income tax (even when it is structured to be equivalent), so that one cannot be confident about how the Court would decide. The proposal for an excise tax in lieu of an income tax does break new ground. A sensible procedure would be to impose the tax initially at a low rate (to minimize possible refunds) and raise the rate if the Court decides to uphold it.

Even if we could get past the constitutional problem, the structure of an excise tax substitute for an income tax would not be easy. In principle, we want to tax only the mark-up added by the public utility firm. If a publicly owned utility that distributes only power buys its power from a private firm for 2 cents per kwh and resells it at an average price of 2.5 cents per kwh, the effect of taxes is already reflected in the 2 cents per kwh, and we are only concerned with the tax related to the additional capital involved in the mark-up. This result might be achieved by a tax rate of 7 percent of sales of electricity (or gas), reduced by the ratio of electricity purchased to electricity sold.

A policy change that removes only part of the discrimination but avoids the complication of a new, intergovernmental tax would be to remove the advantage of tax-exempt financing for publicly owned power companies.

In 1969 Congress denied the tax-exempt bond financing privilege to so-called industrial development bonds. The industrial development bond arrangement was widely recognized as an abuse of a privilege intended to advance governmental functions. The essence of industrial development bond financing was a deal in which a local government bought a facility to be leased to a private company on a long-term basis; the facility would be paid for by municipal tax-free bonds. The bonds were typically based on the lease payments, so that for practical purposes the city was transferring to a private business the opportunity to borrow money at tax-exempt rates. This device had been used in a number of states partly to lure industries away from other areas and was becoming simply an interest subsidy at federal expense to pay for private business plants.

When Congress limited industrial development bond activity in 1969, it enacted several exceptions to the limitation, including one for public utilities—presumably on the ground that public utilities were to some extent a traditional government function. A sensible energy policy now calls for the elimination of this exception to the industrial development bond rule so that the bonds of publicly owned utility companies would sell on the same market basis as the bonds of private utility companies. Arrangements that lead only to lower consumer prices for electricity and gas lead to waste of valuable energy resources.

NOTES TO CHAPTER SEVEN

[1]The specialized tax rules discussed in this section are not applicable to oil and gas pipeline companies.

[2]Testimony of Secretary of the Treasury Douglas Dillon, *Hearings on the Revenue Act 1961*, Committee of Ways and Means, U.S. House of Representatives, and *Hearings on the Revenue Act 1962*, Committee on Finance, U.S. Senate.

[3]Statement of Donald Cook, *Hearings on the Revenue Act 1962*, Committee of Finance, U.S. Senate, pp. 934–890.

[4]Some of this material is H. Averch and L. Johnson, "Behavior of the Firm Under Regulatory Constraint," *American Economic Review*, 52, December 1962 pp. 1052–1069; W. Baumol and A. Klevorik, "Input Choices and Rate of Return Regulation: An Overview of the Discussion," *Bell Journal of Economics and Management Science*, 1. Autumn 1970, pp. 162–190; E.E. Bailey, *Economic Theory of Regulatory Constraint*, Lexington, Mass., D.C. Heath Co., 1973.

[5]F.M. Scherer, *Industrial Market Structure and Economic Performance*, New York, Rand McNally, 1970, p. 531, and R. Spann, "Rate of Return Regulation and Efficiency in Production: An Empirical Test of the Averch-Johnson Thesis," Virginia Polytechnic Institute, Blacksburg, Va., unpublished. Scherer estimates that if the A-J thesis applies, then under plausible assumptions about technology, investment would be carried to a point where its marginal return was about 75 percent of cost. In Scherer's model this implies about 25 percent over-investment. Scherer goes on to qualitative evaluation of industry experience and concludes that the A-J thesis probably holds (pp. 536-537). Spann tests the A-J thesis in a more statistical manner and concludes that over-investment is probably higher.

[6]See *Internal Revenue Code*, Sec. 167(1). The technique is to deny the incentive if the regulatory commission insists on flowing the benefit through in price reduction.

[7]For a rigorous analysis of this point, see Robert Spann's essay in "The Utility Papers," part of the Energy Policy Project series.

[8]The same logic would appear to apply to telephone and telegraph utilities, but we have not fully investigated this matter.

[9]See Chapter 2.

[10]Hellman, *Government Competition in the Electric Utility Industry*, New York, Praeser, 1972, pp. 191, 192. Mr. William McDonald of the Pacific Gas Company has suggested that publicly owned utilities also have some subtle advantages in being treated more generously than privately owned companies by other state regulatory organizations.

[11]A. Wildavsky, *Dixon-Yates: A Study in Power Politics*, New Haven, Yale University Press, 1962, pp. 304–305, 323–325.

[12]Professor Spann supplements this conclusion with a detailed comparison of four more or less comparable electric utilities, two public and two private. See Robert Spann, *op. cit.*

[13]Hellman, *op. cit.*, p. 401.

[14]McCulloch vs. Maryland, 17 *US* 316, (1819).

[15]Graves vs. N.Y. *ex rel.* O'Keefe, 306 *US* 466 (1939).

[16]Alabama vs. King and Boozer, 314 *US* 405 (1938) and N.Y. vs. U.S., 325 *US* 572 (1946).

Chapter Eight

The Conflict Between Environmental and Energy Goals: Taxes or Controls

8.1 THE ENVIRONMENTAL PROBLEM

Many of the short-term manifestations of the energy problem are related to conflicts between environmental policy and energy policy. Some of these are listed below:

(1) Over the last several years it was sensible for oil companies to modify their investment plans for developing new oil reserves in the 48 states to take account of prospective supplies from Alaska. Because of environmental disputes there is still no Alaska pipeline—only a program to build one.

(2) The gasoline shortages of 1973 and 1974 are related in part to lower gasoline efficiency in heavy new-model cars with air-conditioning, automatic transmissions and other power-using extras, plus devices to conform to antipollution regulations.

(3) Over the last decade investment plans of electric utilities have been made in the light of prospective development of nuclear energy. Because of major technical problems and extended delays in administrative and judicial review of whether nuclear plants are in compliance with environmental requirements, this installation is well behind schedule. Installation of fossil plants has been delayed for analogous environmental issues. Fundamental to all this is the simple fact that energy processes account for about three-quarters of air pollution.[1]

These problems are soluble in the long run. There is nothing inherently impossible in dealing with the problems of eliminating pollution from energy processes; it is simply a matter of allocating sufficient resources to the project. The problem is a management one or, more precisely, an economic one: What procedures should we apply to make the decisions about how much energy to produce and how much and how fast to reduce pollution?

119

Eliminating environmental pollution has benefits and it has costs. Much of the long-run energy problem is, at root, a matter of providing for our appetites for energy at reasonable costs. The key to reconciling energy goals and environmental goals is an economic problem of trade-offs between alternative pollution-control techniques and environment goals and the costs of these goals.

Before pursuing this trade-off problem, we need to take account of what might be a communication problem between environmentalists and economists. For many kinds of pollution control, it is possible to identify measurable benefits, that is, benefits such as reduced property or health damages.[2] In many situations, while economists talk about comparing measurable pollution control benefits and costs, environmentalists talk about immeasurable benefits that are difficult to quantify, such as the value of preserving a wild river.

This quantification problem is not an insuperable barrier in economic calculations. We cannot measure the extra satisfaction we get from eating meat, but we can make decisions about how much meat we will buy at various prices. These are basically trade-off decisions. They involve judgments about how much of something else we will give up for meat. Out of a mass of such decisions, we can talk about such measurable things as the price of meat and the demand curve for meat.

In principle, the measurement tools of economists can be applied to immeasurable benefits.[3] The critical question is how much the people involved are willing to give up in order to preserve an environmental benefit. This critical question is not easy to answer. If the issue is developing a large open-pit coal mine in a beautiful valley, who are the people involved? Are they the residents of the valley, the residents around the valley, say, in the same state, or everyone in the United States? The residents of the valley may decide that jobs are more important than the environment. The same question put to everyone in the United States may result in a vote for the environment. The nationwide vote would include people who feel better knowing that there are unspoiled areas in the United States even though they have no current plans to visit there, as well as people for whom a new coal mine would be more competition. Very clearly, the problem of who is involved has wrapped up in it all the issues of federalism (central versus local government) that Americans still argue about.

Even if we know who the groups involved are, it is still a big job to get information on just what people think about trade-offs. Issues usually arise in highly ambiguous political situations. For example, the question might be whether X Company should get a license to mine. Some may prefer to allow X Company to develop the mine rather than have the area undeveloped, but they vote "no" in the hope, often with little evidence, that later Y Company would do a better job of preserving the environment.

Apart from "immeasurable benefits," there are problems with measurable benefits. The main one seems to be the difficulty of isolating effects. In the case of most polluted air, more than pollutant is involved. How damages are attributed to various pollutants may be very arbitrary,[4] even though one pollutant may be much easier to eliminate than another. Particulate pollution in smoke from coal-burning public utilities is far easier to eliminate than sulfur oxides (SOx). Most of our observations about smoke damage to health come from situations in which both particulate matter and sulfur oxides were present.

A critical input to determining rational policy is to measure the marginal-damage function. The term is defined as the incremental damage from a certain type of pollution at various levels of concentration in a region.[5] The marginal-damage function may tell us that a certain concentration of SOx in the air would cause, on the average, an extra $20 per pound of damage; but at a concentration level half that high, the damage from an extra pound might be only $10. Because control policy must involve marginal trade-offs between costs and benefits of pollution control, it is important to know not only the average damage to the environment from a particular type of pollution but also whether the marginal damage rises as the concentration level rises. In the present state of limited knowledge, we can only advise on the basis of current estimates and comment on the kinds of errors that are entailed if these estimates are far off the mark.

Clearly, there is need for the Environmental Protection Agency (EPA) to attach high priority to research on these damage functions.

Since the appearance of acute energy shortages in 1973, various specific environmental regulations, such as those relating to burning high-sulfur coal, have been postponed. This has led to charges by environmentalists that the antipollution program has been sabotaged. A more reasonable interpretation of recent developments is that, due to the acute shortage of energy, the costs of particular improvements in air quality are going to be more expensive than we thought they would be when the regulations were formulated.

It is obvious that an energy shortage could be used by enemies of environmental programs to sabotage them. The best protection against this is to work seriously at the task of structuring environmental programs to involve careful adjustments to benefit/cost comparison. In this framework there is no need for conflict between environmental and energy goals.

8.2   BASIC POLLUTION REDUCTION TECHNIQUES

To reduce pollution, the government has three basic techniques that it can use singly or in combination. These are direct controls, pollution taxes (or effluent fees), and subsidies.

We can reject out of hand a program of primary reliance on subsidies. In the first place, a subsidy for pollution control technology would not be sufficient by itself to cause firms to adopt abatement technology unless the subsidy covered almost 100 percent of the cost. Further, there are a great many pollution control strategies that can be adopted in any one situation; which one will be the most efficient depends on a degree of knowledge of the industry that the government program administrator is not likely to possess. Very likely, therefore, the government would be subsidizing the wrong control techniques.

The most common form of subsidy proposal is a tax credit for pollution control equipment (or rapid amortization of the cost of equipment) that can be readily identified as such. In many cases, however, pollution reduction can be accomplished more efficiently in other ways—by using different fuels, for example—so the usual tax credit subsidizes the wrong technique.

The real choice of technique, then, comes down to pollution taxes or pollution controls. In recent years there has been debate over taxes versus controls. The first problem in deciding the position we want to take on this issue is to get a better handle on the difference between a tax approach and a control approach to making businesses reduce their contribution to pollution. A pure system of controls involves the promulgation of a set of standards or permissible polluting levels, plus a system of prohibitive penalties, usually called fines, for exceeding these levels. A pure tax approach involves a more modest scale of charges levied against the polluter in proportion to the pollution emitted.

In practice, this distinction gets hazy. In a control system the "fine" for exceeding the permissible sulfur oxide (SOx) level could be stated as 50 cents per pound of SOx, which looks like a tax, even though it is high and nearly prohibitive. In a tax system with a 20 cents per pound tax on SOx emissions, an exemption could be provided to cover emission levels too low for the tax to be worth collecting. With this exemption, the arrangement begins to look like a control system. For some firms the "tax" of 20 cents a pound will be prohibitive.

The similiarity of the tax and control systems suggests that much of the difference is in the name, which happens to be a matter of some political importance. If a new proposal is called a tax, it will be referred to the tax committees in Congress; if it is called a control system with fines, it will be referred to the interior committees. Reasonably, an administration wanting a strong antipollution program would want to call its proposal a control system to keep it out of committees less oriented to pollution issues and get it into committees more sympathetic to, or more knowledgeable about, pollution issues.

Despite the vague distinction between the two systems, there is value in analyzing policies to bring out the difference between the tax element and the control element in any technique of inducing business to reduce pollution. The tax element has the feature of imposing some charge on nearly all emissions of the polluting substance. The control element has the feature of imposing a relatively prohibitive charge on pollution activity beyond an allowed level. In the light of this difference, we can make some useful general points before we get into a more detailed analysis.

The popular idea that pollution taxes are "a license to pollute" is misguided. What is at stake is the form of penalty. Controls imply that some pollution is free for the polluter, which does appear to be a license to pollute.[6] Taxes imply a lower financial penalty than controls on some very high pollution levels, which also appears to be a license to pollute. These arguments tend to have an emotional appeal that tells us nothing.

The big problem with controls is setting the critical level beyond which the prohibitive fines will apply. A related problem is that controls are likely to get involved in delaying actions, because whether a firm has exceeded the critical level is likely to be an issue involving administrative and perhaps judicial review. Because of administrative review, for example, the proposed 1975-1976 automobile emission standards have been delayed, and probably no U.S. cars will be manufactured in those model years that meet the original standards. If the rules had been constructed as a tax, it is plausible that some manufacturers would have approached the standard by relying on the price advantage this would give them over competitors who did not meet the standards.[7]

Another problem with controls is that if there are a variety of point sources of pollution (that is, polluting plants), it is likely that the control approach will make total expenditure on abatement technology higher than under a tax approach. Controls tend to result in requirements for uniform reduction in polluting emissions for all firms because tailoring rules to the technology of each firm requires a great deal of knowledge in the controlling organization and, consequently, a large bureaucracy and much opportunity for favoritism. Nevertheless, selective reduction, compared with uniform reductions, can very substantially reduce the total cost. Under a tax approach the reduction is selective. Firms that can reduce pollution easily do so, others pay the tax. In principle, the rate of the tax can be set to bring about any desired level of pollution reduction.[8]

The big problem with a tax approach is that if the marginal-damage function is very steep, that is, if high pollution levels are very damaging and low pollution levels are trivial, the tax does not provide assurance that pollution will not exceed a critical level because it does not, as controls do, slap prohibitive penalties on high pollution levels. In this case, the refinements—such as whether the critical level is set a little higher or a little lower or whether the most efficient technical method of pollution reduction was employed—become trivial compared with the need for a prohibitive sanction on high pollution levels. This is the obvious reason why we have insisted on direct regulation of nuclear pollution while talking about taxes for other industries.

8.3 SOME PRINCIPLES FOR MIXED CONTROL STRATEGIES

Because taxes and controls have different advantages and different drawbacks, tax elements and control elements can and should be combined in the best way to bring about pollution reduction. When it is recognized that the essence of a pollution tax is a uniform charge on all pollution and the essence of controls is a prohibitive charge on pollution over a certain level, it will be clear that the opportunities for a mixed strategy are very wide. A specific pollution "tax" can provide low rates, or no tax at all, for some very low pollution and higher rates on high pollution levels. This is a form of control. There could also be uniform taxes with standby controls to handle emergency situations, or the tax may be suspended in areas where the overall ambient air conditions satisfy the stated environmental quality goals. James Griffin has carefully investigated the characteristics of a good mixed strategy of taxes and controls.[9]

The objective of reconciling our conflicting goals of environmental quality and reduction of energy cost can best be served by making the basic governmental rule fairly uniform over a long period so that firms can adapt their long-run planning to it. The advantages of a pollution tax are lost if the tax rate is highly variable because firms will have to make investment decisions on the basis of the tax outlook at the time of the investment.

Another general issue is regional variation in controls, which has been highly controversial in the debates about environmental policy. Two arguments are advanced against regional differentiation. To the enthusiastic environmentalist, there seems to be something indecent about allowing any degradation of the environment in a region that is already pure. On the other hand, this insistence on extreme purity of the air in presently clean areas will aggravate energy problems.

Any amount of air pollution, say, a concentration of 0.4 ppm of SOx in the ambient air, will do more damage in an urban area than it will in a sparsely settled area.[10] Further, in areas with equal population the added damage from, say, ten more pounds of sulfur oxides will be larger if the concentration level in the air is already 0.3 than if the starting point is 0.1. For these two reasons coal with, say, 2 percent sulfur content should not be burned in New York City where a lot of people breathe air that is already bad. We could be using our energy resources more efficiently, however, if we permit this coal to be used in rural regions where a slight addition to the levels of SOx in the atmosphere will have negligible consequences.[11]

An environmentalist could raise other points against the case for regionalism. We do not really know much about how the marginal damage changes as the concentration level changes, for example. It is conceivable that there are important effects of the first contamination of SOx and that these do not change much with added concentrations. The current Environmental Protection Agency regulations for oil plants levy no penalties on pollution levels that meet both primary and secondary standards and levy progressively heavier sanctions where secondary or both primary and secondary standards are violated. This policy suggests that EPA believes marginal-damage changes rise with concentration levels.[12]

Even if the marginal damage were the same at high and low concentration levels, there is still the contention that it is worse to add pollution to the air that lots of people breathe than to add pollution to the air that few people breathe. The response of the environmentalist to this might be based on a guess that the immeasurable damages to wildlife and vegetation in the less densely populated areas are really as great as the health damages from pollution in urban areas. This does not seem very plausible.

An alternative environmental response might be that people in sparsely settled areas should be entitled to as much improvement in the air they breathe as people in the big city. This does not stand up either. Stack gas cleaners, for example, are an expensive way of handling SOx pollution, and in sparsely settled regions they do not change the air quality for enough people to justify cost. If you really want to help people in that region, you should put the money that otherwise would have gone into the extra cost (over the cost of a cheaper pollution abatement technique) of stack gas cleaners into something more efficient. Better hospitals would undoubtedly do more to improve health conditions in the sparsely populated areas than the small reduction of pollution-related diseases brought about by maintaining very high air quality.

Advocacy of different regional levels of air quality does not mean that we are in favor of deliberately violating rural areas. It means that we favor the philosophy underlying the use of effluent charges, that is, to deal with the conflict between environmental and energy goals by careful cost/benefit comparisons—in this case, region by region.

As we reported earlier, environment-energy conflicts were obvious in the relaxation of environmental regulations after the acute oil shortage in late 1973, which gave rise to charges of betrayal and abandonment of environmental goals. In our view the relaxation was not and need not become a betrayal. It is a recognition that environmental goals are expensive, in terms of other desirable things, such as more energy. This recognition suggests that cost is a constraint on how much environmental improvement we can have; for that reason the exact cost/benefit trade-off characteristic of the effluent charge can help us attain a maximum benefit from environmental improvement for a given cost.

As applied to the regional problem, we can get more use of scarce energy resources if we, say, permit use of high-sulfur coal in areas where the marginal damage from additional sulfur would be negligible. If evidence were presented that the marginal damage from the first speck of atmospheric sulfur in a clean region was very high, it would be consistent with our benefit/cost orientation to incur greater energy costs to avoid this pollution. Our reading of the present evidence is, however, that marginal damages from small amounts of sulfur are very low.

8.4 SOME SPECIFIC PROPOSALS

Our discussion so far has dealt with the general principles of combining the effluent charge strategy with the present regulatory strategy. To reach final judgment about how to apply mixed control strategies, we must examine how they can be worked out in the real world, where it costs money to remove particular kinds of pollution and the removal produces specific benefits. The problem has been examined this way by James Griffin in his study of the control of sulfur oxides.[13]

There are two ways to reduce SOx emissions from burning fuels. We can use fuels with lower sulfur content[14] or we can "scrub" the SOx out of the flue or stack gases during burning.[15]

In the long run, the major sulfur pollution abatement technique will probably be stack gas cleaning, but as of today this technique is just being explored on a pilot basis. The short-run problem is to get as much pollution abatement as possible from using low-sulfur fuels. It is the short-run strategy that has been complicated by the energy crisis in the winter of 1973–1974.

At this point our knowledge of both the benefits and costs of sulfur removal is seriously inadequate. Despite the limitation a systematic look—with the use of a mathematical model—at what we already know about benefits and costs can tell us a great deal. Most important, a model of this sort can tell us which unknowns are critical to the outcome, and it can suggest where we need to concentrate future research.

The critical assumptions that must be made are about the marginal damage associated with more or less sulfur in the air and the technology of still-unproven methods for removing sulfur from stack gases. Griffin's model assumes, first, that an additional pound of sulfur in the air causes marginal damage equal to 29 cents. This figure is used in the marginal-damage estimates released by the Environmental Protection Agency, although Griffin expresses some doubts about its uniform accuracy. He also assumes that stack gas cleaning will be feasible at a price ranging from $3.50 to $6.50 per ton of coal. These numbers work out to a removal cost of 10 cents to 20 cents per pound of sulfur.

The conclusion indicated by the first simple model is that with a marginal damage estimate as high as 29 cents, there seems to be no need for the refinements of an effluent charge. It appears that we would be as well off to require all utilities to install flue-gas desulfurization equipment and be done with it.

There are some complications, however, that make this regulatory solution look less attractive. In the real world things go wrong with complicated machines. Once having complied with the regulation to install the equipment, there is no economic incentive to keep it in good working order. It is probably financially easier for the company to take its time repairing the equipment when it breaks down. (Similarly, there are financial advantages for the automobile driver, namely, gas saving, in disconnecting the antipollution equipment in his car.) The regulatory approach puts the regulator in the nearly impossible position of deciding whether each utility is acting in good faith and making the effort to obtain maximum sulfur reduction. An effluent charge, even when the eventual solution to pollution problems is universal installation of flue gas desulfurization equipment, offers a range of incentives for efficient operation that are very hard to reproduce in a program of exclusive reliance on regulations. Charges on effluent would be a strong supplement to regulations.

A further real-world limitation of the model is that a marginal-damage estimate as high as 29 cents a pound cannot be realistic for all regions. Damage occurs when people and property, as well as wild life and plants, are affected by the SOx. Generally, then, in sparsely settled regions the damage will be much lower. Also, based on the kind of evidence so far offered by the Environmental Protection Agency, there is much doubt that the damage is as high as 29 cents a pound of sulfur even in densely populated regions. Thus, the "estimate of marginal damage" figure may eventually have to be revised downward.

We can extend this discussion of techniques for bringing about pollution reduction to the consideration that regional differences in the pollution problem are quite high. Assuming a general approach to allow for regional difference, it is very likely that in many less densely populated areas of the country the installation of stack gas cleaners would be too expensive to justify when measured against the damage reduction. It might be a completely satisfactory control technique in these areas to continue burning coal with a maximum sulfur content of, say, 2 percent, in lieu of installing flue-gas desulfurization.

The feature of the 1972 Administration proposal making the sulfur tax inapplicable in areas that meet the primary and secondary air quality standards falls short of being a good regional policy. The full implication of accepting the conclusion that sulfur damages are lower in regions where the population density is significantly lower would be to make the rate of sulfur tax differ among regions, or to permit lower air quality standards in less densely populated regions. A convenient way to do this would be to enact a basic federal tax rate and let areas with higher sulfur problems add on a local supplement.

Another difficulty with the Administration's sulfur-tax proposal is that it loses a strong control element by providing for a zero tax rate if the primary and secondary ambient air quality standards are met in a particular region. This approach throws away the valuable feature of a tax that exerts steady pressure on polluting firms to reduce their pollution below the standard level. Judging from the damage estimates that HEW has published to date, there is no solid evidence that the health damages of, say, SOx concentrations of 0.2 ppm in the ambient air are exactly zero if a level of 0.3 meets the primary and secondary standards. It would be beneficial for firms emitting SOx that could reduce their emissions for less than the tax would cost to do so and possibly bring the concentration ratio down to 0.05.

The slow and steady pressure that a pollution tax exerts on a firm to do better than the pollution standard will likely generate some direct benefits, but it will also make it feasible for the region to accept new plants. The Administration's tax proposal, which essentially cuts off pressure for improvement once the standards are met, will have a tendency to encourage firms to operate near the standard. In that case, opening up new plants, even if the new plants follow relatively good pollution abatement practices, will tend to cause the region as a whole to fail to meet the standard; and it will bring on substantial taxes for existing firms.

Under a structure of regional regulations on fuel content, a further element in mixed control strategy that would improve the outcome would be to impose a ·high marginal tax rate on fuels with a higher sulfur content than that specified by the local regulations. Essentially, present regulation procedures provide that a plant be shut down if it has to burn coal with a sulfur content higher than allowed. The provision should be replaced by permission to burn the coal along with a fairly stiff tax penalty for burning it.

This strategy would be especially valuable if by 1976 flue-gas desulfurization has not advanced sufficiently to be in general use and we have high demands for low-sulfur fuels. In this situation direct controls would run into the same sort of technical problem the 1975–1976 automobile exhaust regulations have run into. In the automobile-exhaust case, when EPA decided that the technology was not available, they simply postponed the requirements. In the sulfur case, general reliance on absolute requirements to burn coal with less than 2 percent sulfur content could be stymied by the discovery that there is not enough of this kind of coal available; again, the regulatory requirements would break down because the regulated firms simply could not comply. Instead of a flat prohibition against higher than 2 percent sulfur coals, a 20–cent per pound tax imposed on coal with sulfur content over 2 percent would lead to a rational allocation of what low-sulfur coal we have. The result would be an improvement in air quality that could not be achieved by announcing ambitious goals that could not be enforced. Slow, steady pressure works.

Accenting the tax element in techniques to bring about pollution abatement is valuable in other areas.[16] In strip mining of coal there appears to be a good case for adding a tax element to controls that require land restoration following strip mining. Because land restoration is not always fully satisfactory, there will be local pressures for states to enact prohibitions on strip mining. Then, as some states begin to prohibit strip mining, the increased concentration in the remaining states will aggravate pressures to prohibit it there as well. Political pressures for prohibiting strip mining can be effectively counterbalanced by a tax on strip mining designed to compensate the citizens for the environmental degradation suffered. An even better balance might be achieved if the tax proceeds were earmarked for expenditure on environmental improvement.

The dilemma is that even good strip mining with conscientious efforts to restore the land will involve environmental damages that hurt the people in the area. There will be unsightliness and possibly erosion while the mining goes on, and restoration may not be complete. A control technique with a yes-or-no approach provides no opportunity for bargaining between the value of strip mining and the unavoidable environmental damage. A local-tax-plus-trust-fund approach provides this bargaining possibility. If strip mining is very valuable to the economy, the area will pay the tax. With the money from the tax environmental improvements such as parks, cleanup of waterways and reforestation, can be made, so that local citizens who are concerned about the environment can consider themselves as well of or better off. Without this kind of bargaining possibility, the chances of overly severe repression of strip mining seems considerable.[17]

Similarly, a local tax on petroleum refineries would be an improvement over the present zoning fights that aim to keep out the refineries. The fact that few refineries have been built in the United States recently is one cause of our current oil and gasoline shortages. A local tax could compensate citizens for the environmental damages and offset the predominant local pressure to "put it somewhere else."

In the nuclear field direct controls seem necessary. There is a problem, however, because the remote chance of a catastrophe makes nuclear plants uninsurable. A small tax on nuclear energy could serve as an insurance reserve to cover uninsured damages.

The same sort of insurance function might be served by a tax on offshore drilling. The problem in this case is that litigation involving the varied interests affected by an oil spill is difficult, and there is under-investment in oil spill-abatement efforts. A "tax" in the form of a fine on oil spills with formula schedules for repayment of various injured parties might moderate the loss problems. It could also avoid much of the litigation expense and increase incentives for spill controls.

The fiasco surrounding the 1975–1976 auto emission standards requires some modification of the all-or-nothing character of rules that require a certain standard to be met on new cars. It is quite clear that even if auto manufacturers do not try very hard to develop the necessary technology, the option of closing down Detroit is not feasible. Related to this is the problem that the present system offers no incentive for owner maintenance of an exhaust-control device and that current new-car standards will have the effect of increasing the number of old cars on the road. What we need is a tax on cars that reflects the emission-control standard, which could be enforced through safety inspections.

The basic characteristic of all these proposals is that, under a control device, pollution that gets by the control point is free. It would be rational decision-making to make sure that even this pollution is made expensive. This is the heart of the tax approach. Even when direct controls are to be continued, introducing a tax element would result in cheaper pollution abatement and, if we want it, more pollution abatement.

NOTES TO CHAPTER EIGHT

[1]See James Griffin's essay in *Studies*.

[2]L. Barrett and T. Waddell, *The Cost of Air Pollution Damages: A Status Report,* Department of Health, Education, and Welfare, Public Health Service, Environmental Service, National Air Pollution Control Administration, Raleigh, N.C., July 1970.

[3]Some of the pollution damages are loss of human health or even death, but this does not make measurement impossible. In other areas we might have to choose between more deaths and certain benefits, which implies a measurable benefit for reducing deaths. Highway speed limits of 90 mph, for example, would provide benefits in saving time and involve costs in more deaths. A speed limit of 40 mph would save lives but have high costs in loss of time. Before the energy crisis of 1974, most states were willing to pay for the amount of deaths involved in speed limits of 60 to 70 mph but not the saving of lives involved in 40 mph limits. For a discussion of the value-of-human-life problem see T. Schelling, "The Life You Save May Be Your Own," *Public Expenditure Analysis,* Samuel Chase, Editor, Washington, D.C., Brookings, 1969.

[4]L. Lave and E. Seskin, "Air Pollution and Human Health," *Science,* 169, August 21, 1970, pp. 723–33. Reprinted in *Economics of the Environment,* R. and N. Dorfman, Editors, New York, N.Y., Norton, 1972.

[5]See J. Griffin, *op. cit.,* in *Studies.*

[6]It is not accurate to say that at some low level of SOx emissions a firm is not polluting. Assume that it is consistent with maintaining a safe level of SOx concentration in the air of a region for the firms and householders in that region to emit 100,000 tons of SOx a year. If this much emission takes place and one public utility contributes 50,000 tons of this, it will follow that a new firm that wants to locate in the area and emit 50,000 tons will find that its emission pushes the air quality past the safe level of SOx concentration. One can advocate a policy of favoring established firms over new firms, but it is still clear that the 50,000 tons of SOx emitted by the existing public utility is pollution. It uses up some of the normal capacity of the atmosphere for carrying off wastes, and it makes it harder for new firms to locate in the region to create more local jobs, and so forth.

[7]This is the conclusion of Laurence White, "The Automobile Pollution Muddle," *Public Interest,* Summer 1973.

[8]A study of the water pollution control problems of the Delaware estuary indicates that the cost of raising the dissolved oxygen content by three to four parts per million would be twice as high with uniform treatment as it would be with an effluent charge that would produce differential treatment depending on the river zone and the cost of treatment. See Edwin Johnson, "A Study in the Economics of Water Quality Management," *Water Resources Research* 3, second quarter, 1967, p. 297.

[9]J. Griffin, *op. cit.,* in *Studies.*

[10]Clearly, health damages are proportionate to the number of people, and damages to man-made property are likely to be proportionate to the number of people times their average wealth. Damage to vegetation and animals will be greater in the rural region, but we think that this is the smaller part of damages. Cf. L. Barrett and T. Waddell, *op. cit.*

[11]For a further development of this argument, see J. Griffin, *op. cit.,* in *Studies.*

[12]Barrett and Waddell, *op. cit.,* appear to expect this, pp. 3 and 4, but they find no evidence either way, p. 17.

[13]J. Griffin, *op. cit.,* in *Studies.*

[14]Among major fuels natural gas has the lowest sulfur content. Most crude oils have low sulfur, and where sulfur is present it can usually be refined out. Most coals have high sulfur, and removal before burning is difficult.

[15]One promising technique is to pass the gas through limestone. In this process the SO_x will unite with the calcium to produce more tractable sulfates.

[16]These proposals also draw on Griffin's work, *op. cit.,* in *Studies.*

[17]A sort of bargaining occurs now in zoning practice. A builder who is seeking a variance from the master plan recognizes that his program will involve disadvantages for some citizens, hence he submits a plan that includes considerable green space or other community advantage so that local citizens will decide that, on balance, the variance should be allowed. The tax trust fund idea simply increases and regularizes the bargaining.

Chapter Nine

Tax Policies to Modify Energy Consumption Patterns

9.1 THE PROPER FUNCTION OF ENERGY EXCISE TAXES

So far our discussion of energy policies has concentrated on circumstances that work on the processes of energy supply. The energy problem is, however, a matter of the balance between demand and supply, especially shortages and rising prices, which in turn can be dealt with by reducing demand as well as by increasing supply. In the next three chapters, then, we will discuss the tax and subsidy instruments governments have available to change the *demand* for energy.

This chapter deals specifically with how selective excise taxes might be used to modify energy-demand patterns and whether such policies are likely to be efficient. The next chapter discusses the problem of whether the policies are fair, that is, whether they would hurt the poor. (As it turns out, the regressive measures of a particular tax can be readily offset by various devices, so whether we should have the tax is still a question of efficiency.)

In Chapter Eleven we will discuss how the effect of some demand control policies can be strengthened by linking them to other government policies. The money from tax on a fuel in short supply can, for example, be earmarked for research on ways to make our enormous reserves of coal and shale more usable.

We should start this discussion about the efficiency of excise taxes to control energy demand by asking, "Why control energy demand?"

The answer "to eliminate a shortage" is not good enough. In most market situations, including the energy market, we think that shortages eliminate themselves, or rather that people eliminate them without the intrusion of government. Cutting down the demand for something in short supply cuts down the market response of solving the shortage by more output.

When we organize the problem this way, it can be seen immediately that there are two reasons one might advance for insisting on dealing with the shortage by demand-control measures:

(1) The particular demand is inefficient or wrong-headed, and it is important to prevent more resources from going to satisfy this demand.

(2) The market will not generate much output increase to satisfy the demand. The result is that the shortage will be dealt with in the marketplace mostly by a higher price, which consumers might regard as a windfall for the producer because consumers are not getting more output for the higher price.

9.2   A GENERAL ENERGY TAX OR FUEL TAX

One way of controlling demand is a general tax on all forms of energy. The notion that we are exploring here is that energy growth rates are too high, or our society may be too dependent on energy, and that energy demand in all forms should be cut back. A general levy on energy suggests a tax that relates to energy content, such as the BTU's in each energy form.[1] The possibility that the tax might apply differently to, say, coal, oil, or uranium will be dealt with separately.

A general tax on energy would simply not be a very efficient policy. In the first place, it cannot be justified on the basis of the first principle cited above—that overall consumer demand for energy is just wrong-headed and excessive.

Consumer decisions to use energy in all forms are made up of judgments about the usefulness of an extraordinary range of products and activities, including heating and cooling, transportation, labor-saving devices, aluminum, and steel. We concede that in making many decisions the consumer may simply come up with the wrong answer. For example, people might live longer if they walked on short errands instead of driving a car or they might be healthier if they kept their homes at a lower temperature in the winter. But would they, if left to their own devices, make the choice that was best for them? It is also likely that there are other cases in which people do not use enough energy and shorten their lives by using time on chores that could be simplified by the application of energy.

There is little prospect that government can pontificate over the whole range of consumer choices, maintaining that everyone would be better off with less energy. One can say of consumer choice in these matters what Winston Churchill said of democracy: It is a terrible system of government but better than any of the alternatives.

Further, government can deal with specific defects in consumer choice better than it can with massive defects. For instance, it can be argued that present institutions cause consumers to use too much energy in automobile transportation. The situation can be corrected by a higher tax on automobile transportation; it cannot be sensibly dealt with by a general tax on energy.

There are external costs of environmental damages caused by energy generation that consumers do not have to pay for, which would seem on the surface to be a plausible reason for government deciding that consumers, on the whole, use too much energy. There are, however, better ways to deal with environmental pollution problems than a general tax on energy—both pollution taxes and pollution regulations, for example.[2] Either will impose substantial additional costs on particular ways of providing energy and thus raise average energy prices.

These approaches are superior to a general energy tax for two reasons: some energy forms are more polluting than others, and some energy pollution problems are more costly to control than others. Reliance on environmental regulations and pollution taxes will increase costs and prices for various forms of energy in relation to the seriousness and cost of the special pollution problems associated with each form. In turn, the market response to increased prices, which will vary according to the form of energy, will be to cut back more on the particular energy forms that involve the most costly pollution problems.

Another way in which a general energy tax would be counterproductive is that as the costs of environmental control become more conspicuous, it will become clear that people have different ideas about whether we should have more energy or a cleaner environment. A good environmental program will raise energy prices. To raise prices with an arbitrary tax that has no function in increasing supply will work against making the hard decision to pay an added price for clean energy.

While a general energy tax cannot be justified on the grounds of consumer irrationality, neither can it be justified on the basis of inadequate supply response to high demand—our second reason for applying demand-control measures. Society can in one way or another increase the output of energy very substantially and through a variety of channels, thus, if consumers feel that life would be better with more energy, there is no reason to try to prevent our productive machinery from satisfying this preference.

Applying the second principle, however, does bring up the fact that some energy forms are easier to expand in the short run than others, and there may be a role for some version of selective fuel taxes. This line of analysis gets back to some of the ideas discussed in Chapter Four, which dealt with the royalty problem. If the demand for a particular fuel is so high that it creates "shortages," can the market respond to these shortages and fill them?

In one area—increased demands for low-sulfur coal to meet environmental standards—the answer is yes. There are large quantities of such coal. New mines may need to be opened and more money will have to be spent on transportation, but the price increases for low-sulfur coal will serve a function.

The situation is different for natural gas, as is suggested by the sharp debates over deregulation. A program of fairly sudden natural gas price deregulation might lead to the short-run price over-shooting its long-run equilibrium if demand rises faster than supply.[3] In the transition period the price could be higher until supply catches up. The price serves the function of rationing available supply, and this natural process could probably do the job better than a rationing board in Washington. The trouble is that producers may make gains in the interim.

The Nixon Administration proposed a windfall-profits tax on high oil prices in 1974; as proposed, it fell only on sales above a certain price. A fuel tax structured in this way would not really modify consumption. Instead, it would cut off producer windfalls that might be created by existing consumption patterns.

9.3   INCREASED HIGHWAY FUEL TAXES

In at least one important energy-consuming area—fuel used in transportation—we can argue that existing market arrangements cause consumers to use too much energy, and a consumption tax can be designed to improve the situation. The major transportation fuel-user is, of course, the automobile, which consumes four times as much fuel per person per mile as public transportation.

Even though the energy cost of automobile transportation is high the present system of pricing highway services causes people to think that, in most driving decision situations, the highway is cheaper than it really is. Consumers pay for highways through a gasoline tax, which is a fairly uniform charge that applies to various driving conditions. For example, driving to work during rush hours is considerably more expensive from a social standpoint than driving on an open highway free of traffic. The major difference is that the additional crowding involved in commuter traffic when one more car is added means some loss of highway efficiency for all the other cars. Any particular driver, however, evaluates his decision to use the highway only in his own terms—in terms of the time it saves him and of his automobile expenses. He does not consider the slowdown that he imposes on other drivers as a cost.

Economists who have examined this question uniformly favor

variable highway tolls as a way of charging motorists for highway expenses. The tolls would, for instance, be very heavy for passing bottleneck points during crowded hours and probably zero for using highways in times and places where there is no crowding.[4] The public however, has resisted toll booths, presumably because they slow down traffic as well as absorb money, and the Highway Act of 1956 specifically eliminated tolls as a way of paying for highways in the new interstate system. A few older roads that were financed before the interstate system still have highway tolls.

We think that the continued absence of highway tolls results in commuters underestimating the social costs of getting to work by private automobile. This results in over-dependence on automobile commuting and inadequate reliance on mass transportation, which can move people more efficiently and with much less energy outlay.

An ideal solution would be to institute highway tolls at federal, state, and local levels. Because most of the roads involved are technically local streets, an ideal form of bringing about improved pricing of highway services would be a matching grant program from the federal government that would give better treatment to states that raise some highway funds through tolls.

The prospect of shifting to toll financing is low, however, and alternatives are needed. One promising alternative is a parking tax that would parallel the effect of tolls by making it very expensive for a commuter to bring an automobile into the central city for the day. The ideal base of the tax would be daytime parking fees at commercial lots. These would be supplemented by higher parking meter fees for on-street parking and taxes related to the number of free spaces provided business establishments. Higher parking-meter rates should be imposed mainly in the central city; fringe parking near mass-transit terminals should be tax-free. Because it is the parking of cars involved in rush-hour traffic that should be taxed, there could be a small tax or no tax at all on short-time parking between rush hours.

All of these considerations argue strongly for local imposition of parking taxes. The local government can best identify congested areas and congested times and decide how much refinement to build into the system. The politics of local parking taxes is complex, and there is a fairly well organized automobile lobby. Downtown business interests look on parking taxes as favoring suburban shopping. Again, some federal influence might be brought to bear by introducing differentials in local grants under mass transit in favor of cities that introduce parking taxes.

From an administrative standpoint a relatively easy way of increasing the cost of driving an automobile is to raise the federal gasoline tax or—similar in the long run—to reduce highway expenditures without lowering the gasoline tax. The present gasoline tax could be made somewhat effective by disallowing the present deduction from federal income taxes of state gasoline taxes on non-business driving. This deduction tends to make the existing structure of gasoline taxes even more regressive. There is no income tax deduction for personal expenses, and, in principle, the state gasoline "tax" on non-business driving is a personal expense, a payment for roads.

It should also be taken into account, in regard to a possible increase in gasoline taxes, that the United States is one of the few major countries that does not treat gasoline as a net revenue source, as it does tobacco and alcohol. Most European countries rely on gasoline taxes to such an extent that that the price of gasoline is two to three times as high as it is in this country. As should be expected under this pricing arrangement, the consumption of gasoline per capita is about half as high in Europe as it is here, taking into account income differences.[5] A consequence of the heavy gasoline tax is the much greater preference for small cars as well as greater reliance on energy-efficient public transportation.

Despite the foreign precedent, the gasoline tax is a second-or third-best approach to highway financing. It is regressive and it reflects some, but not all, of the social costs of highway use. The argument for a toll related to congestion implies that the cost of making highway services for off-peak driving is very low.[6] Unfortunately, an increased gasoline tax would burden off-peak driving as well as rush-hour driving.

A higher gasoline tax, if adopted, should be extended to trucks as well as to cars. Even if there is not a general increase in gasoline taxes, the truck-use tax should be increased. To some extent, existing pricing practices lead to over-utilization of truck transportation, which occurs because of the structure of classified freight rates maintained by the Interstate Commerce Commission. These rates, generally related to the value of the produce moved, put railroads at a competitive disadvantage by pricing them out of much of the freight traffic and driving the business to less efficient (including less energy-efficient) trucking. The object of a higher truck tax could be achieved alternatively by abolishing or substantially eliminating the existing railroad freight rate regulation.

There is another problem associated with energy use that is similar to the one concerning highways. According to most authorities, the present taxes on commercial aviation impose an acceptable charge for the airway services that government provides. (Whether airplanes are charged appropriately for related air and noise pollution is dealt with under the principles developed for pollution taxes and regulations in Chapter Eight). The present tax law does not, however, impose an appropriate charge on "general" aviation, which includes both business and private planes.[7] This concession is a notorious energy-using activity; the provision of airway services to general aviation at less than cost is an uneconomic subsidy as well as an unwise energy policy.

9.4 INCREASED TAXES ON ENERGY-USING DEVICES

Because the possibility of an increased fuel tax poses problems, the answer might be to impose a different kind of energy tax.

If we greatly increased the gasoline tax, we could be confident that consumers would reduce their consumption of gasoline in the long run. This observation is based on the consumption patterns in Europe, where gasoline is expensive, and on studies done in the United States that show that gasoline consumption varies with local gasoline prices.

The way in which gasoline consumption is reduced is a bit more complicated, but we can approach it more easily by considering two extremes. At one extreme, consumers who have already purchased heavy, gasoline-consuming cars will go on driving them until they wear out; then a fairly large portion of them will be price-conscious enough to buy a small car with good gasoline mileage. Remember that this is an extreme case, so we can say that consumers are not influenced at all by the gasoline tax when it comes to deciding how much they are going to drive a car they already have. The opposite extreme is to assume that a great deal of the response of drivers to a high gasoline tax occurs through a decision on how much to drive a car they already own.

If our first alternative is the better description of how the world really is, that is, if gasoline prices are only important when it comes to buying new cars, then it would follow that there is not much social purpose achieved by increasing the cost of driving for people who have already purchased cars. The way to increase the cost of driving, would be to hit new-car purchasers with a tax related to, say, the car's weight or horsepower, some structure that would, in effect, provide a lower tax for cars with greater gasoline economy. In fact, the tax could be based on the projected gasoline mileage for the new car, as determined by manufacturers according to certain standards. This kind of tax would relate more directly to energy use than one based on weight or horsepower.

The disadvantage of relying on a tax on new cars as a way of getting at transportation energy use, of course, is that it completely ignores the possibility of inducing consumers to make less use of their present cars, which may also be high polluters. It also overlooks the possibility of cutting the use of new cars. Once he has purchased his car, the consumer will have paid the tax. And whether the tax was high or low, he will continue to find it economical to drive the car to work or around the block for a local errand.

The evidence is that about one-quarter of the consumer response to changing gasoline prices takes the short-run form of driving less with a given automobile investment; the long-run decision about what kind of an automobile to buy accounts for the remainder of the consumer response to a gasoline price change.[8] We think that the whole approach of taxing automobile energy consumption by taxing the machine is far less efficient than a tax on the fuel, and we have seen that a fuel tax can be inefficient.

Regarding energy-use taxes in general, it might be feasible to tax the equipment (appliances, machines, etc.) rather than the energy use. In many cases, it is likely that the rate of use of a given machine is not as variable as the rate of use of an automobile. On the whole, however, it seems clear that taxing the equipment rather than the fuel use is inefficient.

We are left, then, with the parking tax and the toll as the best alternatives to over-use of private cars and finding a more equitable freight-rate structure to cut down on truck-use of fuel and highways.

NOTES TO CHAPTER NINE

[1] For an extended technical discussion on how a general energy tax could best be applied of. G. Brannon, *op. cit.,* in *Studies.* The arguments come out fairly conclusively in favor of placing the tax at the fuel producer's level with a preference for establishing rates for general classes of crude oil, coal, etc.

[2] These matters are explored in more detail in Chapter Eight and J. Griffin, *op. cit.,* in *Studies.*

[3] Demand rises because more gas-using equipment is installed, supply because more wells are drilled and connected to pipelines. It is hard to say which process moves faster. This proposal is only advanced as a possibility since we are not studying the whole issue of deregulation. If some opponents of deregulation are concerned about short-run windfall profits, here is something that could be done about them.

[4] There is extensive literature on this. For example, A.A. Walters, "The Theory and Measurement of Private and Social Costs of Highway Congestion," *Econometrica,* 29, October 1961, pp. 676–696; W.S. Vickrey, "Pricing in Urban and Suburban Transport," *American Economic Review,* 53, May 1963, p. 456; H. Mohring and M. Harwits, *Highway Benefits: An Analytical Study,* Evanston, Ill., Northwestern University Press, 1962.

[5] N. Guyol, *Energy in the Perspective of Geography,* Englewood Cliffs, N.J., Prentice Hall, 1971, p. 99.

[6] William Vickrey, "Pricing in Urban and Suburban Transport," *op. cit.,* p. 456.

[7] Jeremy Warford *Public Policy Toward General Aviation,* Washington, D.C., Brookings, 1971.

[8]H. Houthakker and P. Verleger, "The Demand for Gasoline: A Mixed Cross-sectional and Time Series Analysis," Data Resources, Inc., Lexington, Mass., unpublished.

Chapter Ten

Income Distribution Effects of Energy Policy

10.1 HOW TO THINK ABOUT INCOME DISTRIBUTION EFFECTS

Any energy policy will have effects on the distribution of income—between rich and poor, between wages and profits, between Texas, Oklahoma, Louisiana and the rest of the nation. Furthermore, income distribution effects are important.

Many of our policy debates concern what government is doing or should be doing about poverty. Understandably, then, there is concern about whether a particular program would increase prices paid by the poor or whether it would provide more benefits for the rich than for other people. At a political level, these questions may be of even greater importance. A typical voter facing the issue of deregulating natural gas prices is likely to be unimpressed with "academic" debates about whether deregulation would increase energy outputs; instead, he would place primary emphasis on "what it means to me."

On the other hand, income distribution matters ought to be separable in most policy debates. Let us assume that because of market imperfections, the public simply consumes too much energy at current market prices, and that therefore a general energy tax or a general energy price increase—income distribution effects aside—would be a proper policy. Now assume further that each of these policies, an energy tax or a higher energy price, would hurt poor people. Does this combination of outcomes—a good energy policy that is costly to poor people—mean that the policy should not be adopted? The answer, even for one who is very concerned about the poor, has to be "not necessarily." We can do other things to help poor people and still have the advantages that go with adoption of the energy tax.

We wish that the answer to the question could have been a clearcut "no" instead of a mealy-mouthed "not necessarily." The qualifications were anticipated two paragraphs back, when we said that income distribution matters *ought* to be separable; we did not say that they *were* separate.

In general, policies that involve taxes on energy or increased energy prices would be "regressive." They would absorb a larger portion of the income of the poor than of the rich. Some evidence of this regressivity is given in Table 10.1, which shows the percentage distribution of total income by families of different income classes. It also shows the share each income class pays in direct purchases of electricity and of general industrial energy reflected in the prices of consumer goods. Characteristically, the poor's share in energy costs is higher than their share of income. Families with incomes below $4,000 have only 2.2 percent of the income, but they buy 4 to 5 percent of the energy.

While a tax on energy would bear more heavily on poor people than on rich people, the government could adopt many policies, along with a tax on energy, to prevent this effect. It would be possible, for example, to make a cash payment to all families equal to the average amount of energy tax paid by a family with an income of $5,000. This would assure that, on the average, all families with an income of $5,000 would be as well off as they are today and most families with an income below $5,000 would be better off.

A refund of an energy tax or a payment to offset an energy price increase are only a few of the possibilities. It would also be possible to spend the amount that would have gone into such refunds in ways that would help poor people, for example, by increasing welfare payments. There is no need to enter into a debate over whether one method of helping poor people is better than another. The point is that it is possible to offset the aggregate impact of these measures on poor people.

This does not mean that every poor person comes out even. If one were to repeal the percentage depletion allowance for oil and gas, repeal would increase gasoline prices for a poor family that uses an unusually large amount of gasoline—to drive a long way to work, for example. This family is, to some extent, an accidental beneficiary of present tax depletion policy. That some poor families will lose benefits is not a serious drawback of a policy if the total policy does not work to the overall disadvantage of poor people.

The significance of income distribution effects of various energy policies is basically a political problem, and these effects can be dealt with separately. Whether or not they will be is a different question. People who favor, say, an energy tax if it were to provide a refund for poor people might adopt the political strategy of opposing the tax because they do not really believe that Congress would follow up on the refund part of the proposal. People who disapprove of an energy tax might adopt the political strategy of opposing the refund in order to make the energy tax look bad, that is, regressive.

Our treatment of the income distribution problem is designed to facilitate intelligent discussion. To what extent do these policies change income distribution? What techniques are available to offset these effects? Unfortunately, our political system may not permit much intelligent discussion. In the desire for simple answers people will cite the income distribution effects to argue for or against a basic policy without recognizing that if a policy is basically good, its bad income effects can be offset.

10.2 THE INCOME DISTRIBUTION EFFECTS OF ENERGY PROGRAMS

The main distribution effect we care about is the division between rich and poor. A practical way to identify rich and poor is by total income. Statistics on income are available from tax returns, but most people recognize that there is income that for various policy reasons is left off tax returns—for example, social security benefits, half of capital gains, tax-exempt interest, and the like. Recently, in connection with debates on tax reform, work has been done by Pechman and Okner at the Brookings Institution to develop an estimate of the size distribution of total income.[1]

In addition to income, we need to know how much people spend on energy. Among the studies of family budgets by size of family income, the most detailed is based on a large field survey conducted by the Bureau of Labor Statistics in 1960–1961.[2] The survey is, in some respects, out of date and does not contain many of the expenditure breakdowns that are needed to comment on specific kinds of energy-related expenditures; the expenditure data needs to be supplemented with information from a variety of other sources.

Most students of these matters believe that the correct measure of regressivity or progressivity is the ratio of an energy cost to income before tax.[3] A regressive policy means that the lower income groups' share of the cost of the policy is larger than their share of income before tax.

Expenditures on energy generally take a higher percentage of income in lower brackets than in higher ones. The percentage is typically three times higher in the $2,000-to-$4,000 income class than it is in the top one—$50,000 and up. For those families with incomes below $2,000 it is even higher; but this figure is deceptive because it includes many families with abnormally low or negative income in a particular year who are maintaining their normal consumption patterns by drawing on prior income, savings, or future income credit.

Table 10.1.   Distribution of Income and of Major Energy Costs by Income Level, 1972

Adjusted family income class	Percent of families	Percent of family income	Percent of direct electricity & gas purchases	Percent of indirect energy purchases
Below $2,000	5.4	0.4	1.4	1.2
$2,000-4,000	8.2	1.8	4.0	2.6
$4,000-6,000	11.8	4.1	7.2	5.2
$6,000-8,000	10.0	4.9	6.8	5.8
$8,000-10,000	10.2	6.3	7.8	7.7
$10,000-15,000	20.8	17.8	17.6	20.0
15,000-20,000	12.6	15.2	15.0	16.4
$20,000-25,000	10.1	15.8	15.7	16.3
$25,000-50,000	8.9	21.0	19.1	18.2
$50,000 and up	1.9	12.7	5.3	6.3

Source: Based on a distribution of income developed by Pechman and Okner. Cf. Joseph Pechman and Benjamin Okner, "Individual Income Tax Erosion by Income Class," Washington, D.C., Brookings, 1972, p. 14ff; Benjamin Okner, "Constructing a New Data Base From Existing Microdata Sets: The 1966 MERGE File," *Annals of Economic and Social Measurement*, Vol. 1, June 1972. The figures serving distribution of particular energy costs were taken from the *Impact of Financing Anti-Pollution Legislation on the Distribution of Income in the United States, 1973, 1976, and 1980* by Nancy Dorfman, a Report for the Council on Environmental Quality, The Public Interest Economics Center, Washington, D.C., 1973.

The pattern is particularly conspicuous for expenditures on gasoline. The existing tax law, however, provides a partial "rebate" of the state tax on gasoline, ranging from up to 70 percent for tax-payers in the top bracket to 14 percent for taxpayers in the bottom bracket. The "rebate" occurs because state tax on gasoline used for non-business consumption can be deducted from taxable income on federal tax returns. There is no rebate for people whose income is below tax-paying levels and none for people who take the standard deduction. Because the state gasoline tax rate is typically twice the federal rate, deductibility of the state tax is equivalent to removing the full federal gasoline tax for most high-income consumers.

With regard to electricity and natural gas purchased directly by consumers, the distribution of expenditures is even more regressive than the total consumption pattern (see Table 10.1). The complication in this case relates to the peculiarity of "declining block rates" in electricity pricing. Virtually all sellers of electricity charge high rates for the first purchase, say, 10 kilowatt hours (kwh) a month, then sharply reduce the rates as electric consumption increases. The fact that the poor, in effect, pay more for electricity per kwh is a major reason why the proportion of their expenditures on electricity is so high. A tax on electricity expressed as a definite amount per kwh would be less regressive and probably close to proportional. Even better from an income distribution standpoint would be some steps to change the declining block rate system.

Existing state taxes on utility bills are regressive in the standard pattern. Assessment of taxes by state and local governments on public utilities is undoubtedly regressive in the same way. State severance taxes, on the other hand, are probably progressive.

With regard to existing income tax benefits for oil and gas, the net income distribution effects are hard to identify. Broadly, there are four kinds of effects that could result from the tax provisions, each having different income distribution consequences. These are summarized in Table 10.2.

The portion of the tax benefits that went into lower prices would in general, be distributed more in favor of low income consumers than high income consumers. The high income consumers, of course, would enjoy far more absolute benefit from the lower energy price.

Nearly half of the tax benefits that are reflected in higher royalties go to high income people. Because a royalty income has a tax benefit attached that depends on the marginal tax bracket of the "royalist," these properties are typically sold to rich people.[4] Thus, producer incentives, to the extent that they raise royalty rates, increase inequality of income distribution.

Table 10.2.   Income Distribution Consequences of Income Tax Benefit Provisions for Minerals

Type of effect	*Income result*
1. Increase royalties	To a large extent, helps high incomes, but smaller part comes back to government in off-shore leases, etc.
2. Reduce product prices especially gasoline	Helps low incomes
3. Lead to excessive drilling (higher cost)	Reduces real GNP, probably proportionately
4. Increase profits	Helps high incomes

By and large, government programs to subsidize research on extraction of better energy forms from coal and shale would have some of the same effect, but in the long run the more important effect would probably be to lower energy prices and help those with low incomes.

One other group of government energy-related programs has major income redistribution effects—programs for reducing environmental pollution. To talk of income redistribution in this area, though, we need to know how the benefits of a cleaner environment are distributed, and so far there has been no solid work on the benefit-distribution effect. One could guess that homeowners who are well off already live some distance from industrial pollution sources, and that poor people live on the low-value land near factory noise and smoke. On the other hand, preservation of wilderness areas and recreational water areas must be of primary value to people with resources for country homes or camping trips. Overall, there are tough conceptual problems about how one values health benefits, recreational benefits, or even intangible aesthetic benefits for a poor family and a rich family respectively.

Even though we have little knowledge of the income distribution of the benefits of environmental improvement, we do have information about the distribution of costs of environmental improvement. They will be about the same as with any energy tax and will not change much whether the environmental program is based on taxes or regulations. The cost burden will be relatively heavier on lower incomes.

10.3 INSTRUMENTS FOR CHANGING THE INCOME DISTRIBUTION EFFECTS OF ENERGY POLICY

As we said in the beginning of this chapter, government has at its disposal a wealth of tools to offset any unfavorable income distribution effects of a proposed energy policy. The problem is largely a political one of putting together two program elements.

One way to offset the regressivity of an increase in the gasoline tax would be through an increase in welfare payments that would compensate the poor for the additional gasoline tax burden. But let us look at the political aspects. If a welfare increase were to come to the vote, there would be the arguments that typically arise in welfare debates—whether larger welfare payments increase illegitimacy, whether welfare mothers cheat, and so forth. In addition, people inclined to support the gasoline tax increase on the condition that its regressivity would be offset would have considerable reason to fear that the offsetting increase in welfare payments might be defeated after the gasoline tax was approved.

The problem is even more difficult in the long run. It may be that this year's increase in welfare payments specifically intended to offset the gasoline tax burden would be looked at next year as simply part of the welfare payment schedule. The result might be that some future increase, say, for cost of living, would be lower than it should have been by the amount of the increase added this year. Because of the political complexities, there would seem to be advantages in dealing with the regressivity of a gasoline tax by some device closely linked to the gasoline tax itself that would not get lost in welfare debates.

Despite the drawbacks, there are two important advantages in relying on specific antipoverty programs as a way of dealing with income distribution problems. One advantage is that a good program to deal with poverty could be designed so that it includes very serious efforts to identify the poor who are to be helped. Existing welfare programs try to do this, although they do not do it very well, and large portions of the poverty population are not taken into account. Some of the recently proposed versions of a negative income tax, principally, the Family Assistance Plan (FAP), are basically efforts to find a more systematic way to identify the poor who should be helped. In contrast, an antipoverty program that is much more closely related to an increase in gasoline taxes is the proposal to use the proceeds of the increased gasoline tax, through an earmarked trust fund, to subsidize mass transportation. On the whole, subsidized mass transportation would help poor people. It would also help a lot of non-poor people, and it would fail to help a lot of poor people who were not conveniently located to use the busses or subways.

The other basic advantage of relying on systematic poverty programs to meet the income needs of the poor concerns the subtle problem of implicit marginal tax rates. This problem has come to be recognized much more clearly as a result of the recent public discussion about the negative income tax and FAP. Whatever benefits are given to poor people on the condition that they will be withdrawn when those people get enough other income to be considered no longer poor, the withdrawal of benefits acts just as any other tax rate on income does. If the combined explicit and implicit tax rates are too high, they could be a strong disincentive to earning income.

The implicit tax rate problem was most conspicuous in welfare programs that sought to provide some minimum level of subsistence in a way that reduced the amount of the welfare payment dollar for dollar by any income that the client family obtained from other sources, such as part-time employment. This constitutes an implicit 100 percent tax on wages that would raise the client family out of poverty. (A major objective of the negative income tax proposal is to reduce the marginal tax rate to, say, 50 percent.)

As welfare programs were examined in the light of how they actually operated, it was evident that the present structure of multiple, uncoordinated antipoverty programs implied a host of marginal tax rates that could add up to *more* than 100 percent. Poor families in subsidized public housing, say, would lose the subsidy if their income rose. Those obtaining free day care for children would lose the benefit if their income rose. Others obtaining food stamps would lose them if their income rose. All of these losses of benefits constitute additional marginal tax rates. Together they add up to a sort of welfare trap. The family on welfare might find that efforts to become self-supporting would do very little to improve its consumption levels and could even make them lower.

As we look toward the adoption of special devices closely related to energy programs that are designed to help poor people, we are quite likely to find that some aspects of the programs, while they may be politically more effective, have substantial inefficiencies as relief for the poor.

The important problem of obtaining a close relationship between the energy program and the program to offset its heavy burden on the poor makes it imperative that we explore a number of devices that could be built into the energy program itself. Recognition that these devices are likely to be inefficient ways to help the poor suggests two bits of advice. We should, in the first place, closely examine these programs to determine how much they benefit the poor and how much they benefit people who are well off. We should also be suspicious of programs that provide a benefit to poor people that is withdrawn as income rises. This advice seems rather contradictory, because not withdrawing a benefit when income increases means that some of the benefits will go to the non-poor; but it is the price that must be paid in dealing with second-best devices to help the poor.

If the federal tax on gasoline were increased by a relatively small amount, a simple way of offsetting the regressive impact would be to repeal the federal income tax provision for deduction of state gasoline taxes. This would be the equivalent of almost a 4–cent increase in the gasoline tax for most high-bracket taxpayers, and it would be a negligible increase for low-bracket taxpayers because most of them use the standard deduction. If the increase in the gasoline tax were in the neighborhood of 7 cents to 10 cents, however, the added regressivity of the tax would more than offset the progressive impact of the repeal of the state gasoline tax deduction.

A more direct offset to regressivity would be to provide a refundable tax credit equivalent to the average increase per family in gasoline tax or price at some modest income level, say, $4,000 or $5,000. With a 5–cent increase in the gasoline tax or price, this would involve a credit of about $20 per family. In aggregate, the tax credit would offset about one-quarter of the five billion dollar impact of a 5–cent gasoline tax or price increase. For people filing tax returns, this sort of credit is very easy to handle; and most poor families who do not have taxable income still file tax returns to obtain refunds on wage withholding. Other poor families are most likely to have contact with government through welfare or social security or unemployment insurance, and so it would be relatively easy to provide a vehicle for creating the refund for people who are not familiar with filing income tax returns.

It is noteworthy that this refund device need not be phased out as income gets higher. Phase-outs run into the old problem of increasing the marginal tax rate. Furthermore, the major objective of a higher tax on gasoline would be to modify consumer choices in the direction of modes of transportation that are more economical in their use of energy, and this objective could be achieved even if the entire revenue from the added gasoline tax were refunded. We could, for example, design a schedule of income tax credits equal to the average additional gasoline tax paid in each income bracket. The total credits would equal the additional tax collections. Each family would be entitled to the credit even if they consumed no gasoline, so they could improve their financial position by using less gasoline. This sort of refund device could apply to any of the energy taxes that we will refer to later.

If a general tax were imposed on the use of electricity by households, the tax could be made less regressive by basing it on the quantity of electricity, or natural gas, used rather than on the size of the electric bill. It could be made more progressive still by providing that the federal tax would apply progressively on the amount of the electric bill. This could be handled most simply by a monthly exemption plus a flat rate on the balance, so there would be several rate brackets depending on the size of the bill. In view of the prevailing pattern of declining block rates in the charge for electricity, this device would tend to offset the inherent regressivity of the present electricity price system; it would also be economically efficient in discouraging the marginal use of electricity that only appears to be low cost because of the existing rate schedule. The proposal would apply somewhat erratically to apartment units, which are not individually metered, but such units presumably get special benefits already from the block rate schedule, so that there should not be any overall unfairness. If a utility already used some kind of average-consumption calculation in applying its rate schedule to a group of tenants, the same average calculation could be applied with the progressive rate. A substantial portion of the electricity costs of consumers are hidden in their payments for goods and services, the production of which required purchases of electricity. There is no convenient way to make a general tax on electricity use progressive except through the refund device described in our discussion of the gasoline tax.

The costs of pollution control in the energy industry, which are shifted forward into energy prices, will have a regressive effect on income distribution because of the typical pattern of energy purchases. One way to deal with the problem would be to provide direct federal subsidies for polluting firms to meet part of the cost of pollution control. From a broad economic standpoint, though, this is very inefficient because it tends to increase the demand for the products that involve heavy environmental pollution.

There is one part of the pollution cost, however, that might be efficiently subsidized—the cost of research related to pollution control methods. Successful research on energy pollution control methods provides benefits for all firms having that pollution problem, but unless the firm paying for the research is able to recapture these benefits completely through the sale of patent rights, the firm is not likely to push research outlays to the efficient level. This situation, along with the basically regressive character of pollution costs, suggests that the federal government should not only subsidize, through cash payments, some research on pollution control, but it also take a very liberal attitude on what constitutes research. Subsidies might well be extended to pilot operations.

A tax on energy-using appliances is somewhat less regressive than a tax on the energy itself. This is particularly true in the case of the automobile, where the outlays on automobile purchases, including used-car purchases, are moderately progressive through most of the income scale. Unfortunately, an automobile tax, even one measured by horsepower or vehicle weight or rated miles per gallon, will likely be a highly inefficient tax from an energy control standpoint; so, on balance, it appears to be a poor way of offsetting the regressivity problem in a tax on gasoline.

NOTES TO CHAPTER TEN

[1]See Table 10.1 and the note on source.

[2]U.S. Department of Labor, Bureau of Labor Statistics, *Survey of Consumer Expenditures, 1960-1961.*

[3]The essay by Professors Davie and Duncombe in *Studies* deals in more detail with some issues of methodology.

Chapter Eleven

Earmarked Revenues for Energy Projects

11.1 TWO KINDS OF EARMARKING[1]

In the current discussions about energy policy, there are many who would favor the establishment of an energy trust fund. It is usually proposed that the revenue from a general tax on energy go into the trust fund for energy-related projects, such as research on new energy sources or research on pollution control related to energy and scoring of reserves of energy resources.

Frequently, proposals of this sort are advanced by people enormously concerned about possible energy shortages and rising energy costs. In one view, the problem is seen as a possibility that Congress will not be sufficiently concerned about energy problems in year-to-year budget operations. For these people, a trust fund holds promise of obtaining a one-time commitment to large energy programs and effectively putting energy outside the annual grabbing of shares in a tight budget.

The prospect that Congress would surrender annual control on energy budgets is quite remote. In most trust funds Congress maintains annual appropriations; this could be expected in the energy field. Energy decisions are important, and they are not likely to be put off. In addition, energy decisions have enormous impact on various pressure groups, which makes these decisions subject to political decisions. Congress is likely to insist on a close review of projects that could have important ramifications on, say, the price of oil or coal or the income of oil states or coal states. Finally, the concern of Congress about budget control as evidenced by the establishment of a joint committee on budget procedures in 1973 is an additional indication that the legislature will not completely delegate expenditure authority on energy matters.

Even though a trust fund is not likely to provide a huge commitment of money beyond annual budget review, the fund can have an important influence on budgetary decisions. Congress has, for example, retained annual authorization and appropriation control over the highway trust fund. The highway trust fund, and the highway user taxes that go into it, nevertheless, constitute a commitment to an annual program of highway expenditures in the area of the present federal highway program of five billion dollars a year.

The implication of all of this is that proposals to establish some sort of earmarked fund for energy purposes relate to some subtle influences on expenditure policy. Analyzing these subtleties is fairly complicated. There is also a common view that both public finance and political specialists are uniformly opposed to the use of earmarked revenues. This is not an adequate description of the public finance viewpoint, but it is a good place to begin exploring the problem.

The public finance viewpoint that has no patience with earmarked revenues grows out of the notion that government expenditure decisions should be based on a cost-benefit type of analysis. A devoted systems-budget man would recognize, of course, that given the present state of the art it is not possible to quantify all of the costs and benefits of alternative programs. He is usually optimistic, however, that he can quantify a number of costs and benefits and that he can make good judgments about the non-quantifiable ones; any kind of influence on the budget decision growing out of a prior commitment to trust funds, he would reason, can only clutter up the rational decision process. He would also assert, that the need for outlays on items such as energy research should be examined each year in the light of all of the evidence available. It may be that by next year the state of research on new energy sources would be so far along that additional outlays would not be required or that the state of research is such that some specific jobs have to be finished before we know where to go next, it might not be profitable for instance, to spend much more money.

This is, of course, a highly optimistic view of systems budgeting, and it seems to be rather out of touch with a good deal of the political tugging and hauling that is characteristic of budget decisions in the real world. In reality the areas in which the budget-maker does not have good estimates of costs and benefits are very broad, and most observers are not convinced that his judgments about non-quantifiable costs and benefits are any better than anybody else's.

Once it is conceded that an unhindered technical approach to budget-making is not likely to be very satisfactory, it becomes sensible to explore the potential of other possible influences on the expenditure-decision process, namely, the earmarked fund.

There are two conceptions of how an earmarked fund can contribute to a better outcome of the budget-decision process. One we call pure decisional earmarking. The feature of this approach is that it attempts to provide an opportunity for the public, that is, the voters, to express an attitude toward particular expenditures, in this case, energy-related expenditures. In this approach it is unimportant whether the tax earmarked for energy purposes is an efficient user charge. The important thing is that the public has an opportunity to express a political view on a particular expenditure program as distinct from other things on which the government spends money.

A citizen who feels that the government is already spending (1) too little on energy problems and (2) too much on defense and welfare would be inclined to express to his congressman strong opposition to an increase in the income tax because the bulk of the revenue raised by income taxes goes to pay the costs of defense and social welfare programs. As for a general tax on energy to be spent for energy purposes, the citizen who opposed an increase in income tax should express a favorable view toward an energy tax on the ground that he wants government to deal with the energy crisis even if he is opposed to more expenditure on defense and welfare.

A different kind of earmarking, which we call functional earmarking, is one in which the budget-maker looks at the receipts under the earmarked tax as a measure of the benefits from the expenditures that the earmarked taxes finance and uses this as a guide to how much should be spent on those categories.

It is extremely difficult, for example, to make judgments about the benefits to the public associated with improvements in national parks or national monuments. An obvious supplement to making a budget decision is to impose an admission charge for national parks and national monuments and to use the proceeds as a guide to public interest in various kinds of parks and monuments in various locations. In functional earmarking, the important consideration is not the political process in which the earmarked tax is voted, but the individual decisions that citizens make when they choose to pay the tax by visiting a particular park or a particular monument.

11.2 AN ENERGY TRUST FUND AS DECISIONAL EARMARKING

There have been proposals for a trust fund for research on new energy resources, energy-resource reserves and reduction of pollution associated with energy, and for stockpiles of energy resources, such as oil. The proposals are that these trust funds be financed by the proceeds of a general tax on energy or particular taxes on energy pollution or a tariff on imported oil and gas.

In this analysis of decisional earmarking, we will put aside the questions of how closely the particular tax source is related to the energy expenditure objectives in order to examine the political issue of a tax increase related to energy expenditures. Let us say, for example, that there is a proposal to impose a 3 percent surcharge on all income tax liabilities, with the proceeds going into an energy trust fund. Removing the income tax surcharge from the basic income tax itself creates a separate political issue about which citizens can express a preference.

When we look at the energy expenditure problem this way, we can see that energy expenditures are a different kind of issue from defense and welfare. Feeling runs high on defense and welfare spending. Some might be concerned about militarism and feel that a very high level of defense outlays will lead to arms races and military adventures. Others might be concerned that excessively generous levels of social welfare programs will lead to an undermining of the work ethic.

It is doubtful that this same degree of active opposition would apply to higher levels of expenditure on energy programs. At any ambitious level of expenditure on government energy programs some citizens would be in favor and some opposed, but we think the opposition would be different from the opposition to defense and welfare. One might feel that additional outlays for research on atomic energy or solar energy or coal gasification are not likely to be very useful without feeling that the outlays pose the same sort of threat that welfare or defense outlays do. The public would probably be more amenable to a tax increase specifically labeled for energy purposes than a general tax increase that would be available for defense and welfare.

We conclude that outlays for energy-related programs would be higher under a decisional earmarking trust fund financed by something called an energy tax.

11.3 FUNCTIONAL DECISION EARMARKING FOR ENERGY

There is still the question of functional earmarking. Are there some potential taxes that could be employed as user taxes related to energy—that is, can consumer decisions to buy the taxed product (to pay user taxes) be a useful guide to government expenditures?

At first glance, it might appear that a general tax on energy would serve this function. As people spent more on energy, the tax receipts would go up; this would serve as a signal that there might be greater public benefits in government programs to increase energy availability or lower energy costs.

On careful analysis, though, the logic does not stand up well because government expenditures in the energy field are not likely to apply uniformly across the whole energy industry but are more likely to have strong impacts on particular segments. This is most striking in the area of research and development; the burden of the tax would probably fall on the current major energy sources, such as oil and gas, while the proceeds would be used to develop oil and gas substitutes.

In research and development fields it would seem more sensible to associate taxes and expenditures with a view toward the future. Expenditures for research and development might well be decided by analyzing prospective research benefits, particularly the external or public benefits associated with research. If a particular research effort was highly successful, for example research on gasification of coal, the demand for and the price of coal would probably rise substantially, and owners of coal deposits would enjoy considerable windfalls. The appropriate connection between expenditures and taxes would be to impose, say, a severance tax some time in the future if it should turn out that government policies have so changed the market situation for an energy resource that the owners were enjoying substantial windfall profits. Unlike a general energy tax, this kind of tax would not discourage the development of new energy forms because it would only be imposed on new energy forms as they became successful; the magnitude of the tax would be related to windfall profits.

A more promising kind of functional earmarking arises in connection with government expenditure related to conserving energy resources and, in particular, the national security argument for building up some protection, such as an inventory reserve, for possible interferences with overseas supplies. The security problem is related primarily to the specific demand for oil and gas, particularly oil. Dependence on overseas supply is reduced to the extent that we can make greater use of electricity from coal and nuclear energy, and it would not be sensible to tax these coal and nuclear resources to build up a petroleum reserve. It makes more sense to impose a specific tax on petroleum and on natural gas if imported liquified natural gas becomes a major source of energy. The thrust of this tax would be to impose on petroleum users the cost of ensuring petroleum supply, which would be related to their decision to use petroleum. (An interruption of petroleum imports would not be a problem for the military establishment because domestic supplies are more than adequate for this.) Chapter Five discusses the two devices of an import tax and a strategic inventory. Earmarking one for the other would be logical.

Related to the conservation problem is the special situation that we referred to in Chapter Ten: Because of the peculiarity of our present method of paying for highways—the gasoline tax—consumers are induced to excessive use of automobiles and under-use of mass transit. In the absence of general reliance on variable highway tolls, an increased gasoline tax earmarked for additional outlays on mass transit would be efficient. The highway problem is primarily the traffic jams associated with commuting to and from work; simply increasing the gasoline tax until mass transit becomes relatively attractive involves too heavy a charge for highway use at times when there is no crowding. A combined program of increasing the cost of driving along with reducing the cost of mass transit would be a more efficient way to economize on the use of energy in transportation. This combination lends itself to earmarking.

The other major category of government expenditure in the energy field relates to programs involved with pollution control. By and large, these programs do not appear to be good prospects for functional earmarking. As is made clear in the general discussion of energy-related pollution problems in Chapter Eight, the important function of pollution regulations or pollution taxes is to associate environmental costs with consumer demands. It is essentially counterproductive for the government to subsidize selective polluting activities by paying for part of the pollution costs. An exception could be made for government subsidies to firms for research and development on the grounds that research and development work is likely to have benefits on firms other than the ones who pay for it. Without the initiative of government subsidies it is likely that private firms would underinvest in research and development, though this would not be unique to pollution-control programs. Under a general program for subsidizing research and development, the projects related to pollution control could be evaluated on their prospective public benefits, the same grounds on which any other research and development work is evaluated.

NOTES TO CHAPTER ELEVEN

[1]The argument of this entire chapter is developed more extensively in my essay in *Studies.*

Chapter Twelve

Policies on Imported Oil

12.1 OIL IMPORTS AND THE VALUE OF THE DOLLAR

In Chapter Five we discussed policy on oil imports as it related to protecting domestic energy producers. We turn now to consideration of oil imports as a foreign trade problem that arises when the United States is a heavy purchaser of foreign oil. The problem involves the impact of our import program on the balance of payments and the cartel of petroleum exporting countries (OPEC). This section deals with the balance of payments.

The prospect, as of early 1973, that the United States would substantially increase its oil imports led to some dramatic forecasting. John McLean of Continental Oil, for example, told the House Ways and Means Committee:

> A fuel deficit of $20 billion [his estimate for the early 1980's] will impose a well nigh intolerable burden on our trade position and make it increasingly difficult to maintain the stability of the dollar in world financial markets.[1]

At that time the world oil price was about the same as the United States price. Because the world price more than doubled in late 1973, one would hesitate to make any forecast of imports for 1980. To import a given number of barrels will cost more at the current price, but the higher price will mean that we will import fewer barrels and make more use of domestic resources. President Nixon even talks about U.S. self-sufficiency in energy, which suggests no oil imports.

Despite Administration resolves, whether we become self-sufficient in energy will be largely a matter of relative price differences between domestic energy and oil and gas imports. If OPEC pushes the world oil price to $15 a barrel, it will be easy for the United States to be self-sufficient. If the world price drops to $4, self-sufficiency would not only be harder to attain but also not worth attaining (this was discussed in Chapter Five). It emphasizes the role of prices to point out that another way of describing U.S. self-sufficiency is to refer to it as "OPEC pricing itself out of the U.S. market."

Because we cannot predict the OPEC price, it is useful to address the question of whether there are special balance-of-payments problems for the United States if we spend a large amount on oil imports. Mr. McLean's 20 billion dollar figure for the "fuel deficit" is as good as any. If the OPEC price falls, we will undoubtedly make many imports. With a higher price there would be fewer barrels imported, but we might still end up with a 20 billion dollar deficit. With a still higher price we could become self-sufficient and not need an import policy. If we do need an import policy, though, what should it be?

Even if the United States does make high expenditures on oil imports, we think that Mr. McLean's forecast is overly dramatic, because even in the arcane world of international trade and foreign exchange there are adjustment mechanisms. A country can, over time, greatly increase (or greatly decrease) its imports of certain commodities. Prices and exchange rates change as necessary, and the world economy rolls along.

In international trade, the most valuable part of the adjustment mechanism is a reasonably flexible exchange rate. It appears from recent experience that this is the way chronic imbalance of international payments will be adjusted.

Adjustment is not without cost. Instead of referring to this as a "well nigh intolerable burden," the sensible procedure would be to evaluate the cost involved in the United States's adjusting itself to a substantial increase in the level of oil imports. One striking fact is that by 1980 the level of our imports and exports should be approaching 200 billion dollars a year. When we look at the financial problem of greatly increased oil imports in relation to the international payments process, we see that a 20 billion dollar deficit in one commodity could easily be swallowed up by surpluses in other commodities or by foreign investment earnings.

Whatever the overall situation is, the point can still be made that, all things considered, the dollar will be stronger in international markets if we import less oil rather than more. This proposition holds up even if by 1980 the world value of the dollar is higher than it is today (which we think will be the case).

The issue here is really a technical one: What are the costs to the United States of adjustments necessary to accommodate a high rather than a low level of oil imports? The issue is not some threat that at a particular level of oil imports the United States will become bankrupt, or otherwise have "intolerable burdens." In fact, adjustment costs are relatively independent of whether the total U.S. dollar position is weak or strong.[2]

The essential point is, as we said, that other things being equal, the U.S. balance of payments will be weaker—or less strong—if oil imports are higher. The most general adjustment mechanism is a change in the exhange rate. In practice, it is likely that small disruptions in the balance of payments will be ignored or offset by temporary policies. Nevertheless, it is appropriate to estimate the effect of changes in the trade balance as if they were accommodated by exchange-rate adjustments. This should provide an estimate of the cost of various methods of adjustment. Alternative levels of oil imports, therefore, have some costs to the United States over and above the immediate payment for the oil because they imply a different exchange rate for the dollar. This, in turn, means that with significantly higher oil imports the dollar exchange rate, that is, the value of the dollar, should be lower. As a consequence, we will be paying more for imports than we would have otherwise, and we will be getting less for exports than we would have. Paying more and getting less are referred to as "terms of trade" effects.

There are three characteristics of the world trading and investment system that are crucial to estimating the significance of the terms-of-trade effect.[3] The first is the portion of oil receipts that come back to the United States at given exchange rates. This can happen in a variety of ways:

1. repatriation of foreign profits after tax of U.S.-owned oil subsidiaries;
2. repatriation of foreign profits of U.S.-owned shipping companies;
3. proceeds of oil sales being used by foreign nationals or foreign governments to buy U.S. goods, or U.S. securities or proceeds being spent for the purchase of other goods or securities from those who will use the proceeds to buy U.S. goods or securities.

In principle, the calculation of the "back-flow" coefficient associated with additional oil imports would call for a complex model of world trade and investment flows, including financial investment. There is no detailed model available, but there have been several recent attempts to estimate some of the relevant components of back-flow.[4] In the various studies estimates are that the sum of these back-flows is on the order of 40 to 50 percent of the dollar payment for U.S. oil imports. The portion of OPEC receipts respent on U.S. goods or capital is commonly 20 percent[5] to 33 percent.[6] In addition, the profit return could be on the order of five to 10 percent and the shipping income 10 percent.

Given a back-flow estimate of 50 percent, the decision to buy more imported oil in 1980 would imply a terms-of-trade loss to the United States through exchange rate adjustments; the magnitude of the loss would depend on the elasticity of demand for imports and exports. By and large, the currently available estimates of these elasticities are quite low, in the neighborhood of 1.5. If we assume that the elasticities are as high as 2, it would follow that the real cost to the United States of importing five billion dollars more oil would be about 7.5 billion dollars. Alternatively, we would be as well off to spend 7.5 billion dollars on oil substitutes from U.S. sources.[7]

This argument substantially supplements the case made in Chapter Five for dealing with the oil security problem by imposing a tax on imports in an amount necessary to provide a strategic inventory of crude oil reserves against a future interruption of supply. If that inventory cost were relatively high, say about $1 per barrel, we would need an import tax of close to 30 percent at the 1973 price of crude oil, which is in the range of our terms-of-trade analysis that suggests imports involve real costs of 7.5 billion dollars compared with money costs of five billion dollars. If one were inclined to estimate the feedback effect at less than 50 percent, or the import and export elasticities at less than 2, a case could be made for a tariff of higher than $1 a barrel.

In Chapter Five the import tax/inventory proposal was suggested as a policy that might become relevant in the future if the world oil price moves down to the point where we become heavily dependent on oil imports from the cartel. We also argued that the import tax/inventory approach was superior in that it provided security for the combination of tax incentives plus import quotas that was the United States policy from the mid—1950's to the early 1970's.

Because of the balance-of-payments effects of reliance on imports for our oil needs, there is an extra cost associated with large imports. An import tax/inventory system would be particularly relevant because it would reflect the extra cost of imports. We think that the general argument against tariffs—that they lead to trade wars—is not a controlling factor here because we are already engaged in a sort of economic war with a cartel devoted to acquiring as much of our money as it can in payment for its oil.

12.2 OTHER POLICIES WITH REGARD TO OPEC

Apart from the relatively technical financial consequences of paying for imported oil, it is clear that the whole character of the future world oil market depends on the relative strength of the producer nations (OPEC) in relation to the consumer nations. This point was brought home by the worldwide crisis induced by producer embargos, production cutbacks, and price increases in late 1973.

In 1969 the U.S. Cabinet Task Force on Oil Imports projected a cost of imported oil of about $2 a barrel through the 1970's.[8] A series of renegotiations of oil-concession agreements involving the Persian Gulf countries and Libya brought about increases in the tax-royalty payments to the governments in those countries that were passed on in higher prices. At the time of this writing the latest increase brought the tax-royalty payment to Saudi Arabia to $7 per barrel and the price in the Persian Gulf to $7.50 to $8[9] or $8.50 to $9 delivered on the East Coast of the U.S.

It is very clear that in a market where the producing firm can make a profit with a total margin of 50 cents per barrel, the most significant feature of the market is the set of circumstances that protects the $7 take that the producing countries can exact. In the market for international oil the "set of circumstances" is a coalition of the principal oil exporting countries—a coalition that can effectively control the behavior of a fairly limited number of large producing companies by means of what is in effect an excise tax. Although the tax is accepted as an income tax for U.S. foreign tax credit purposes, its important feature from the standpoint of the cartel is that is precludes an operating company from deciding to follow a price-cutting policy. Companies have a tax liability set by posted prices and formula costs per barrel. They can charge higher prices, and they can at any time compete by cutting prices within the margin of price over tax, but the tax acts as a floor price. In recent years the floor has been raised sharply, in effect exerting the cartel's power, which is beyond the abilities of even the powerful oil "giants" to control.

The history of cartels, however, leads one to think that they do not have great staying power, as witness the common resort to governmental authority to protect the cartel.[10] Basically, the cartel members are likely to have divergent interests. In the case of oil the long-run interests of the cartel countries are likely to vary according to different ratios of current production and reserves, different demands for oil, different time discounts, and the like.

The remarkable thing about the precipitous oil price increase engineered by the OPEC in late 1973 is that it has not been accompanied by dramatic production cutbacks. There have been announcements of cutbacks from the September rate of production by some countries, then a partial lifting of the cutbacks. For example, the Persian Gulf countries, which account for about two-thirds of the non-U.S., non-communist world output, announced a 22½ percent cutback in November, along with specific embargos. In December a further announced cutback of 5 percent was rescinded and a 10 percent increase substituted. Iraq, the North African countries, Venezuela, and Indonesia have apparently not made much of a cutback. World output would normally have been 5 percent higher in early 1974 than a year before; and the only cutback—now ended—was in the Persian Gulf. The world production level was temporarily lowered by only 15 percent—a very small reduction to support a tripling of the crude price.[11]

It is plausible that not much of a cutback was necessary, because in the short run the demand for energy in particular forms is very inelastic. This results from the simple engineering fact that machines are almost always designed to use a certain type of fuel and in the short run the possibilities of using less of the fuel are very limited. The user has little choice but to pay the price demanded. In the long run, however, fuel users can adapt. Oil-burning machines can be replaced by coal-burning machines. Technology to convert coal to a liquid fuel can be advanced. Oil exploration and development in the North Sea or on the U.S. outer continental shelf can be pushed ahead.

It seems clear that the OPEC has not yet established the discipline necessary to sustain a long-term world oil price in the $8 to $9 range. If production continues at current rates, an inventory will be accumulated. There is uncertainty whether the OPEC can establish production cutbacks then, or whether there will be sales at lower prices undercutting the cartel.

There is ample reason for the consuming countries to consider whether their energy policies strengthen or weaken the OPEC cartel. This sort of concern was reflected in the conference of consumer nations held in Washington in February of 1974.

An exhaustive analysis of strategy *vis a vis* the cartel goes beyond the confines of this book. We do, however, offer some comments on how tax and subsidy policy can contribute to weakening the cartel. In a sense, the OPEC situation is so dominant as a cause of our present energy problems that the question "How does it affect the OPEC situation?" may become the acid test of any domestic policy.

This test would, for example, argue very strongly against the resumption of oil import quotas. Basically, a quota is an announcement that the importing country will not be receptive to price reductions in the usual market sense of increasing purchases at lower prices. One could hardly imagine a policy that works more to the advantage of a cartel than an announcement by a customer that there will be no advantage to a cartel member who tries to break the cartel discipline in order to make more sales at lower prices. There should be no consideration of a restoration of quotas.

Other implications of appropriate strategies for the United States as a consuming nation facing a producer cartel are developed by Davidson, Falk, and Lee.[12] One circumstance in the oil business that is relevant to this strategy analysis is the role of "user" cost. A fundamental consideration for an owner of oil or gas property is that, if he makes an agreement to license production this year, he reduces his ability to license production in the future. The user cost to the landowner of selling oil now is the prospective loss of future sales. Thus, the expected future prices of oil have a great deal to do with the willingness to increase current oil production. If we were able, for example, to develop atomic energy to the point where it would be far more efficient than fossil fuels in the gneration of electricity and through lower electric prices would increase the feasibility of an electric automobile, one would expect that the price of oil in 1990 would be much lower than it is now. If the OPEC countries believed this forecast, the user cost of production would fall drastically. Their strategy would be to sell more oil now before the oil market collapses.

This consideration suggests that the United States should continue a policy of investing considerable amounts in new energy technologies, including coal liquefication and gasification, shale processing and nuclear energy—both fusion and fission. An equivalent program would be to push ahead with a relatively generous leasing policy with regard to the development of the oil and gas reserves under the U.S. outer continental shelf.

The development of resources in the outer continental shelf will involve some environmental problems. At the same time, the United States has been able to exact very high lease bonus payments and should be willing to trade some potential lease bonus payments for requirements that the producers spend additional amounts on environmental protection. With environmental protection the U.S. has more to gain from increasing the production from the outer continental shelf reserves than it has from maximizing its royalty rate or lease bonus payment in order to improve the appearance of the short-term budget. The same sort of argument applies with respect to other new energy sources.

The strategy of pushing rapid development of known oil supplies has apparently been decided upon by the United Kingdom, where policies are now directed at early establishment of high production rates from the North Sea deposits.[13]

Rapid development has some complications. If one were convinced that the world faced an imminent danger of running out of energy reserves, such as that described by the Club of Rome,[14] then a rapid-development strategy could cause owners of existing oil deposits in the producer countries to expect that while the consumer countries might "artificially" push down energy prices during the 1980's by, say, the year 2000, prices would be even higher. This line of analysis suggests several approaches:

(1) A major aspect of our development strategy should be to push energy forms that will not "run out," expecially nuclear and solar energy.

(2) As long as we have believable, even if not assured, energy prospects to which we can turn if fossil fuels run out, it would be extremely risky for any producer country to hold back its oil production in an anticipation of future price increase.

(3) A rapid development strategy is likely to be expensive for the countries that are presently OPEC customers. Development must be pushed in the face of a continuous threat that the world oil price could drop and the substitute-energy-source industries could be high-cost producers in competition with cheap oil.

As we argued earlier, the possibility of reverting to an import quota is an inefficient tool for protecting new energy industries that might grow under a rapid development strategy. A more sensible device would be a contingency reserve strategy, such as that outlined in Chapter Five. An important feature of the reserve strategy is that it holds the promise of being able to reward particular OPEC countries that are willing to cut the price of oil. This would be most valuable in the next few years, when a break in the OPEC price might come early enough to permit us to cut back on our long-run program of developing alternative sources of energy.

A more explicit strategy toward the OPEC was discussed in Chapter Six. The prospect of the OPEC controlling production, and hence price, is weakened when more countries become oil producers. This consideration argues generally for the kind of incentives granted by the present tax law, which allows a deduction for intangible drilling expenses. As cited in Chapter Six, the allowance amounts to a subsidy from the U.S. Treasury to an international oil company going into a new country. We think that the objective of this incentive is sound but that the present tax provision is a poor way to go about it because the tax provision becomes more valuable the less it is needed. It would be more efficient to commit the Treasury to some sort of guaranteed subsidy for losses in the event a new venture is unsuccessful.

Another long-run circumstance that is, relevant to the strength of the cartel is the present "convenient" situation in which the producer countries can, in effect, use the oil companies as "tax collectors." (The term is Adelman's.) In any cartel a continuing problem is to keep members of the cartel from shaving prices in particular transactions to obtain short-run advantages, such as a higher share in the cartel output. In the case of international oil, because the cartel is effectively run by the governments of the major producing countries, it is convenient to use the tax law as a bulwark for the minimum price. Even these countries tax laws are harder to change than prices. If the producer countries actually took over ownership of oil production, it would be in their interest to continue to employ the existing bureaucracy of the oil companies to serve as production managers and technicians. The big difference would be that the countries would then be in the position of setting prices and would not have the floor of taxes to insure price discipline against particular countries attracted by the prospect of a larger sale at a lower price.

The view that nationalization of the oil companies would, in the long run, weaken the cartel is expressed by Adelman and is also attributed by Adelman to Sheik Yamani,[15] although the enthusiasm of the host countries to take over control of the producing companies seems to be unabated, judging by almost continuous steps in that direction. It is probably beyond the ability of the United States to change this pattern of nationalization, and there is little reason to try.

NOTES TO CHAPTER TWELVE

[1]*General Tax Reform,* Panel Discussion before the Committee on Ways and Means, U.S. House of Representatives, February 26, 1973, p. 1,244.

[2]One may point out that from 1969 to 1972 the foreign exchange markets were rather chaotic. There did not seem to be a smooth adjustment to the U.S. deficit problem and the ensuing devaluation. We think that this pattern is not applicable to the oil case because (1) from 1969 to 1972 we were just beginning to break away from the rigid dollar value that had developed under the Bretton Woods system; (2) current international monetary talks will lead to a more orderly method for changing exchange rates; and (3) the short-term pressures from 1969 to 1972 were quite exceptional, in part due to the long period of an over-valued dollar.

[3]The following analysis follows the argument developed in more detail in Davidson, Falk, and Lee, *op. cit.,* in *Studies.*

[4]Helmut Frank and Donald Well, "U.S. Oil Imports: Implications for the Balance of Payments," *Natural Resources Journal,* July 13, 1973, pp 431–447, and "Energy and the Balance of Payments," Research and Planning Staff, Domestic and International Business Administration, U.S. Department of Commerce, October 18, 1973, reprinted in *Fiscal Policy and the Energy Crisis,* Committee on Finance, U.S. Senate, November 20, 1973. The major thrust of the model developed by the Department of Commerce was to estimate the total impact on the U.S. balance of payments of all developments in the energy field, including, for example, additional U.S. exports because of OPEC countries respending their earnings from oil sales to Europe and Japan. Our present concern, however, is solely with how the U.S. balance of payments depends on U.S. oil imports.

[5]"Energy and the Balance of Payments," Department of Commerce, *op cit.,* p. 72.

[6]*The Oil Import Question,* Cabinet Task Force on Oil Import Policy, Washington, D.C., Government Printing Office, 1970. See also Frank and Well, *op cit.*

[7]A detailed derivation of this result is given in Davidson, Falk, and Lee, *op. cit.*

[8]*The Oil Import Question, op. cit.,* p. 40.

[9]Testimony of Exxon Corporation before the House Ways and Means Committee, February 6, 1974.

[10]Cf. Scherer, *Industrial Markets Structure and Economic Performance, Rand McNally, 1970, pp. 158–164;* also *Stocking and Watkins, Cartels in Action,* New York, Twentieth Century Fund, 1946.

[11]This summary is based on various reports in *The Oil and Gas Journal,* especially December 31, 1973.

[12]*Op. cit.,* in *Studies.*

[13]See "North Sea Investment Seen Totalling $50 Billion" *Oil and Gas Journal,* December 10, 1973.

[14]Donella Meadows, *The Limits of Growth,* New York, N.Y., Universe Books, 1972.

[15]M.A. Adelman, *op. cit.,* pp. 215, 261.

Summary

This study explores the role of taxes and subsidies as instruments of government policy to bring about a better response in the economy to the set of problems we call the enrgy crisis.

The work assumes that tax and subsidy policy must be decided in relation to the strengths and weaknesses of the market price system. Ultimately, decisions on how to solve energy problems must be made on the basis of: (1) the value to consumers of marginal increases in the energy supply; and (2) the cost of producing these increases in energy supply. The best way of getting this information is through responses of consumers and producers to market prices.

The market price system, however, has various defects, which if ignored can lead to bad energy decisions. Some of these defects can be offset by judicious use of taxes and subsidies. One example of a defect of energy market prices is that the market does not take into account pollution effects; ignoring the effect of pollution, in turn, makes energy too cheap and leads to excessive use of it. Another defect is that when foreign oil prices are below U.S. prices, the market does not reward activities that increase the national security of the United States.

We accept the conclusion of another study in this series that energy markets are broadly competitive. Even if the monopoly power of energy companies is stronger than we think, there is not much that tax and subsidy policy can do about it; we need, instead, to have stronger antitrust policy.

Looking at taxes and subsidies as ways of dealing with market defects leads to a number of important conclusions:

(1) The existing tax provisions to encourage production of natural resources for energy (percentage depletion and the current de-

171

duction of intangible drilling and mine development expenses) work very unevenly among various energy sources; for instance, the tax benefits are equivalent to 13 percent of the market price for oil and gas, but only 1 percent for coal and nothing for hydro or solar energy. This is a basic defect of the depletion concept that mistakenly provides a greater income for value added by exploiting valuable natural resources than for value added by manufacturing. If we have to distinguish between these two kinds of income, it would be sensible to tax income from exploiting natural resources more heavily than income from manufacturing. (See Chapters Two and Three)

(2) The effect of these tax benefits in the past has been divided this way: about 50 percent to lower market prices, about 40 percent to higher royalties, and maybe 10 percent to higher company profits. In the light of the present price situation of energy materials, these subsidies are likely to benefit royalty recipients even more, because the price will tend to be fixed by the price of imported oil. The total annual revenue cost of these tax benefits for energy resources is currently about 4 billion dollars. (Chapters Two and Three)

(3) The special risk features and the relatively high capital needs of the oil and gas industry do not justify the special benefits of percentage depletion and deduction of intangible drilling expenses for successful wells and development costs for mines. Allowing deduction of dry hole costs is the appropriate treatment for the peculiar risks of exploratory drilling. (Chapter Four)

(4) World oil prices were, until a few years ago, much lower than U.S. prices; despite the current highs, we could have low prices again (if the OPEC cartel breaks apart). Low world oil prices can lead to importing half or more of our oil needs, and this dependence on imports is a potential national security problem. Import quotas, such as those we had in the 1950's, are a bad way to deal with such a situation. It would be better to rely on a tariff, plus a strategic reserve inventory of oil. (Chapters Five and Twelve)

(5) The thrust of the points listed above is that for the immediate future the United States should accept the inevitability of a higher domestic price for energy, especially oil. The higher price will provide uniform market incentives for all energy forms, which is far better than relying on tax rules that heavily favor oil and gas. A high price for energy will also cause users to economize on energy use. (Chapters One and Five)

(6) A high price for energy will give oil and gas producers and the recipients of oil and gas royalties much higher incomes, which provides the opportunity to remove obsolete tax benefits without hurting the industry. The increased Treasury revenues will make it possible to transfer income to energy consumers to prevent higher future energy prices from burdening the poor. (Chapters Five and Ten)

(7)   U.S. oil companies conduct large-scale foreign operations in the production of crude oil and natural gas as well as related foreign activities. The profits from production are heavily taxed by the governments of foreign countries where production takes place. It is consistent with general principles of our tax law that there be no further U.S. tax on foreign income fully taxed abroad. The current means of achieving this—the foreign tax credit—goes beyond relieving double taxation of production income; it results in inappropriate tax relief on other foreign activities. These tax provisions relating to foreign income should be tightened up. (Chapter Six)

(8)   The major problem with public utility taxation is the tax benefit available to publicly owned electric and gas distribution systems. The benefit arises from absence of income tax on the return on the capital invested and access to tax-exempt bond financing. Removing these two benefits would remove a subsidy that encourages electricity and gas use that, in turn, aggravates the energy shortage. (Chapter Seven)

(9)   There is little to be gained from a general tax on energy. Excise taxes on energy use should be concentrated on those energy uses that involve social costs not paid by the user. A major user of energy that involves high social cost is the private automobile, which should be more heavily taxed. The gasoline tax is an inefficient way of taxing the auto use with the highest social cost—commuter traffic. Both highway tolls at congestion points or downtown parking taxes would be better auto user taxes. (Chapter Nine)

(10)   The other energy uses with high social cost are those involving polluting fuels, especially high-sulfur fuels. Congress has rejected economically efficient pollution taxes in favor of a complex regulatory system that will probably prove very unsatisfactory when it has to be enforced in tough cases. It is not likely that we will change the general orientation of anti-pollution policy from primary reliance on direct regulation. We should, however, introduce several pollution-tax elements into the present system (that is, utilize a mixed control strategy) to improve the enforcement outlook. (Chapter Eight)

(11)   It would be bad social policy to reject policies involving higher market prices for energy or higher taxes on energy on the grounds that these measures "hurt poor people." Maintaining a lower final price for energy directly helps those who are well off far more than it helps the poor, because people with higher incomes spend more on energy. (Both the well-off and the poor are hurt by the indirect effects of artificially low energy prices.) The sensible way to deal with the burden of higher energy prices or energy taxes on poor people is to transfer income to the poor. This can be done by such broad measures as higher welfare payments or relief from social security payroll taxes, or it can be done by more specific measures, such as a refundable "energy-price tax credit." (Chapter Ten)

(12) It is also sensible policy to subsidize research in new energy sources. If these programs are expensive, it is likely that the outlays will be larger if energy programs are financed through a special trust fund. Where subsidized research makes some fuel resources very profitable (through higher royalties), government should try to recapture some of its advances through special severance taxes. (Chapter Eleven)

Index

About the Authors

Gerard M. Brannon, currently Research Professor at Georgetown University, has been involved with the formulation of tax policy for almost thirty years. He has served as an economist with the Joint Congressional Committee on Internal Revenue Taxation, U.S. Bureau of the Budget, and the House Ways and Means Committee; more recently he was Director of the Office of Tax Analysis, Office of the Secretary of the Treasury.

Professor Brannon's advanced degrees include both an M.P.A. and Ph.D. from Harvard University. He has published numerous articles in the major journals of finance and taxation.